S0-BOA-665

Math through the Ages

Math through the Ages

A Gentle History
for Teachers and Others

Second Edition

William P. Berlinghoff
Fernando Q. Gouvêa
Colby College

Oxton House Publishers
Farmington, ME
2014

Oxton House Publishers, LLC
P. O. Box 209
Farmington, Maine 04938

phone: 1-800-539-7323
fax: 1-207-779-0623
www.oxtonhouse.com

Copyright © 2002, 2014 by William P. Berlinghoff and Fernando Q. Gouvêa. All rights reserved.

No part of this publication may be copied, reproduced, stored in a retrieval system, or transmitted in any form or by any means, electronic, mechanical, photocopying, recording, or otherwise, without written permission of the publisher. Send all permission requests to Oxton House Publishers at the address above.

Printed in the United States of America

20 19 18 17 16 15 14   10 9 8 7 6 5 4 3 2 1

Publisher's Cataloging-in-Publication
(Provided by Quality Books, Inc.)

Berlinghoff, William P.
  Math through the ages : a gentle history for teachers
and others / William P. Berlinghoff, Fernando Q. Gouvêa,
Colby College. -- Second edition
    pages cm
    Includes bibliographical references and index.
    ISBN-13: 978-1-881929-54-3
    ISBN-10: 1-881929-54-X

  1. Mathematics--History.   I. Gouvêa, Fernando Q.
(Fernando Quadros) II. Title.

QA21.B47 2014   510.9
                        QBI14-600158

Preface to the Second Edition

In the dozen years between the initial appearance of this book and today, two ongoing events have motivated us to prepare this revised edition. The first is the significant expansion of published scholarship in the history of mathematics, including many new books accessible to non-specialists. The second is the gratifying widespread use of this book, both here and abroad, and the continuing demand for it. In addition, several more topics, admittedly chosen by personal taste, seemed to merit sketches of their own.

Five new historical Sketches — on the tangent function, logarithms, conic sections, irrational numbers, and the derivative — have been added. We opted not to change the numbering of the original Sketches, so the new ones are numbers 26–30. Also new is the "When They Lived" section just before the Bibliography. It replaces all the parenthetical birth/death dates that had appeared somewhat intrusively (and, alas, inconsistently) throughout the text, after the name of each mathematician or other prominent historical figure.

Parts of the "Nutshell" overview have been rewritten extensively to reflect recent scholarship. The "Books You Ought to Read" section and the Bibliography have been thoroughly reworked to reflect recent publications. The index has also been expanded and improved, and the "For a Closer Look" piece at the end of each Sketch has been revised to account for sources that have appeared in the past decade or so.

In preparing this revision we have tried to be mindful of the many college professors whose courses have been built, at least in part, around the first edition. To that end, Sketches 1–25 have remained substantially the same. To be sure, parts of some have been reworked a bit. In particular, there have been significant adjustments in Sketches 1, 5, 9, 15, 17, 18, and especially 19. Nevertheless, the "story lines" of all the Sketches are unchanged, so they are still compatible with the Questions and Projects of the first Expanded Edition. An Expanded Second Edition with Questions and Projects for the new Sketches is in the works.

Maine, August 2014

Preface to the First Edition

This book grew out of a few casual hallway conversations in the Colby College Mathematics Department about two years ago, but its roots are much deeper and older than that. For many years we have been interested in the history of mathematics, both for its own sake and as an aid in teaching mathematical concepts to a wide range of audiences. One of us has used it as a major ingredient in several college mathematics texts for liberal arts students and as an important part of his contributions to an NCTM Standards-based high school mathematics series. The other has done considerable background research in the field, has participated in the Mathematical Association of America's *Institute for the History of Mathematics and its use in Teaching*, and teaches a course in the history of mathematics at Colby. We are convinced that knowing the history of a mathematical concept or technique leads to a deeper, richer understanding of the concept or technique itself.

Unfortunately for teachers and other people with some interest in mathematical history but relatively little time to pursue it, most books on the subject are dauntingly large. If you want some historical background as you prepare to teach quadratic equations or negative numbers, or if you are just curious about the history of π or the metric system or zero, where would you look? The indexes of most history books will point you to a disjointed scattering of pages, leaving to you the task of piecing together a coherent picture. A topical search on the Internet is likely to inundate you with information, some reliable, some spurious, with little guidance as to which is which.

We decided to write a book with your needs in mind. The main part of this book is a collection of twenty-five short historical sketches about some common ideas of basic mathematics. These sketches illustrate the origins of an idea, process, or topic, sometimes connecting seemingly distinct things that share common historical roots. They are preceded by a brief panorama of the history of mathematics, from its earliest days to the present. This provides a skeletal framework of important people and events that shaped the mathematics we know today, and it supplies a unifying context for the separate, self-contained sketches.

Of course, the choice of sketch topics was quite subjective; we were guided partly by our own interests and partly by our sense of what might interest teachers and students of mathematics. If you would like to suggest a sketch topic for the next edition of this book, we invite you to submit it to Oxton House Publishers, either by mail to the address on the copyright page of this book or by e-mail to one of the authors.

We have made every effort to reflect accurately the historical facts as they are known today. Nevertheless, history is far from an exact science, and incomplete or conflicting sources often lead to conflicting judgments of fact among scholars. Some stories about mathematical people and events have evolved over many years, creating a body of "folklore" with very little hard documentary evidence to support it. Despite their potential to annoy historical scholars, many of these stories — like folk tales in every culture — are valuable, either as allegories or as mnemonic "hooks" to help you (or your students) remember a mathematical idea. Rather than ignore such anecdotes entirely and lose their value, we have opted to include some of the more interesting ones, along with appropriate cautions against taking them too literally.

To help you track down more information about any topic that interests you, the section entitled "What to Read Next" is an annotated list for further reading. It includes some pointers to reference works, but its heart is a short "ought-to-read" list of books that we think anyone interested in the history of mathematics probably would enjoy.

A note about notation: In recent years, some history books have been using B.C.E. ("before the common era") and C.E. ("the common era") in place of the more traditional B.C. and A.D., respectively. Depending on which historian one consults, this is either (a) the notation of the future for historical literature, or (b) a passing "politically correct" fad. Without taking a position on this question, we have opted for the notation that we believe to be more familiar to most of our potential readers.

Acknowledgments

We are indebted to many colleagues from near and far for sharing their knowledge and for their forbearance in responding to our sometimes peculiar questions. In particular, we thank mathematics education consultant Sharon Fadden in Vermont, Jim Kearns of Lynnfield High School in Massachusetts, and Bryan Morgan of Oxford Hills Comprehensive High School in Maine for reading and commenting on earlier versions of the book. Special thanks also to Georgia Tobin for creating the TEX symbols for Egyptian and Babylonian numerals, and to Michael Vulis for converting them to PostScript format; to Robert Washburn of Southern Connecticut State University for providing some of the material in Sketch 6; and to Eleanor Robson, who generously gave us permission to use one of her drawings of Old Babylonian tablets (on page 63).

We are deeply grateful that one of us was able to participate for two summers in the MAA's *Institute for the History of Mathematics and its use in Teaching.* IHMT helped to transform a lifelong interest in the history of mathematics into a solid base of knowledge on which it was easy to build. Special thanks to IHMT organizers Fred Rickey, Victor Katz, and Steven Schot, to the Mathematical Association of America, its sponsoring organization, and to all the IHMT colleagues — an interesting, varied, knowledgeable, and helpful bunch of people. Many of them answered questions and made useful suggestions while this book was being written, earning still more of our gratitude in the process.

Our debt to the many historians of mathematics whose work we read and used as we were writing this book is enormous. Were it not for the giants on whose shoulders we attempted to stand, we couldn't possibly have done the job. We have tried, in bibliographical notes scattered throughout the book and in the "What to Read Next" section, to point our readers towards some of their work.

We also would like to thank Don Albers, Martin Davis, David Fowler, Julio Gonzalez Cabillon, Victor Hill, Heinz Lüneburg, Kim Plofker, Eleanor Robson, Gary Stoudt, Rebekka Struik, and the members of the *Historia Mathematica* group for their answers to our questions. Of course, any mistakes that remain are our own.

Contents

History in the Mathematics Classroom

W here did mathematics come from? Has arithmetic always worked the way you learned it in school? Could it work any other way? Who thought up all those rules of algebra, and why did they do it? What about the facts and proofs of geometry?

Mathematics is an ongoing human endeavor, like literature, physics, art, economics, or music. It has a past and a future, as well as a present. The mathematics we learn and use today is in many ways very different from the mathematics of 1000, or 500, or even 100 years ago. In the 21st century it will no doubt evolve further. Learning about math is like getting to know another person. The more you know of someone's past, the better able you are to understand and interact with him or her now and in the future.

To learn mathematics well at any level, you need to understand the relevant questions before you can expect the answers to make sense. Understanding a question often depends on knowing the history of an idea. Where did it come from? Why is or was it important? Who wanted the answer and what did they want it for? Each stage in the development of mathematics builds on what has come before. Each contributor to that development was (or is) a person with a past and a point of view. How and why they thought about what they did is often a critical ingredient in understanding their contribution.

To teach mathematics well at any level, you need to help your students see the underlying questions and thought patterns that knit the details together. This attention to such questions and patterns is a hallmark of the best curricula for school mathematics. It is the driving force behind the Standards for Mathematical Practice, a major component in all levels of the Common Core State Standards. It is also reflected in the National Research Council's "Framework for K-12 Science Education" section of their 2013 report, *The Mathematical Sciences in 2025* [32]. Most students, especially in the early grades, are naturally curious about where things come from. With your help, that curiosity can lead them to make sense of the mathematical processes they need to know.

So what's a good way to use history in the math classroom? The first answer that comes to mind is "storytelling" — *historical anecdotes*, or, more generally, biographical information. Here's a typical scene. When introducing the idea of how to sum an arithmetic progression, the teacher tells a story about Carl Friedrich Gauss.

> When he was about 10 years old (some versions of the story say 7), Gauss's teacher gave the class a long assignment, apparently to carve out some peace and quiet for himself. The assignment was to add all the numbers from 1 to 100. The class started working away on their slates, but young Gauss simply wrote 5050 on his slate and said "There it is." The astonished teacher assumed Gauss had simply guessed, and, not knowing the right answer himself, told Gauss to keep quiet until the others were done, and then they would see who was right. To his surprise, the answer that the others got was also 5050, showing that young Gauss was correct. How had he done it?

Telling such a story achieves some useful things. It is, after all, an interesting story in which a student is the hero and outwits his teacher. That in itself will probably interest students, and perhaps they will remember it. Being fixed in their memory, the story can serve as a peg on which a mathematical idea can hang (in this case the method for summing arithmetic progressions). Like most biographical comments, the story also reminds students that there are real people behind the mathematics that they learn, that someone had to discover the formulas and come up with the ideas. Finally, especially when told as above, the story can lead the class towards discovering the formula for themselves.

But this example also raises some questions. That story appears in many different sources, with all sorts of variations. The sum is sometimes another, more complicated, arithmetic progression. The foolishness of the teacher is sometimes accentuated by including elaborate accounts of his reaction to Gauss's display of attitude. Many, but not all, versions include an account of Gauss's method. Such variations raise doubts about the story. Did it really happen? How do we know? Does it matter?

To some extent, it doesn't matter, but one might feel a little queasy telling students something that might not quite be true. In the case of our example, it's actually not hard to settle at least some of these questions. The story was told to his friends by Gauss himself when he was older. There is no particular reason to doubt its truth, though it's possible that it grew in the telling, as stories that old men tell about themselves often do. The original version seems to have mentioned an unspecified arithmetic progression that involved much larger numbers, but overall the account above is likely not too far off the mark. Unfortunately, it is not always easy to find out whether an anecdote is true. So, when using an anecdote, it's probably a good idea to make some sort of verbal gesture to suggest to students that what they are hearing may not necessarily be the strict historical truth.

The main limitation of using historical and biographical anecdotes, however, is that too often they are only distantly connected to the mathematics. This book, while including some such stories,[1] hopes to point you toward some other ways of using history in the classroom, ways that more tightly intertwine the history with the mathematics.

One way to do this is to use history to provide a *broad overview*. It is all too common for students to experience school mathematics as a random collection of unrelated bits of information. But that is not how mathematics actually gets created. People do things for a reason, and their work typically builds on previous work in a vast cross-generational collaboration. Historical information often allows us to share this "big picture" with students. It also often serves to explain *why* certain ideas were developed. For example, Sketch 17, on complex numbers, explains why mathematicians were led to invent this new kind of number that initially seems so strange to students.

Most mathematics arises from trying to solve problems. Often the crucial insights come from crossing boundaries and making connections between subjects. Part of the "big picture" is the very fact that these links between different parts of mathematics exist. Paying attention to history is a way of being aware of these links, and using history in class can help students become aware of them.

History often helps by adding *context*. Mathematics, after all, is a cultural product. It is created by people in a particular time and place, and it is often affected by that context. Knowing more about this helps us understand how mathematics fits in with other human activities. The idea that numbers originally may have been developed to allow governments to keep track of data such as food production may not help us learn arithmetic, but it does embed arithmetic in a meaningful context right from the beginning. It also makes us think of the roles mathematics still plays in running governments. Collecting statistical data, for example, is something that governments still do!

Knowing the history of an idea can often lead to *deeper understanding*, both for us and for our students. Consider, for example, the history of negative numbers (see Sketch 5 for the details). For a

---

[1] Several sources for stories of this kind appear in "What to Read Next."

long time after the basic ideas about negative numbers were discovered, mathematicians still found them difficult to deal with. The problem was not so much that they didn't understand the formal rules for how to operate with such numbers; rather, they had trouble with the concept itself and with how to interpret those formal rules in a meaningful way. Understanding this helps us understand (and empathize with) the difficulties students might have. Knowing how these difficulties were resolved historically can also point out a way to help students overcome these roadblocks for themselves.

History is also a good source of *student activities*. It can be as simple as asking students to research the life of a mathematician, or as elaborate as a project that seeks to lead students to reconstruct the historical path that led to a mathematical breakthrough. At times, it can involve having (older) students try to read original sources. These are all ways of increasing student ownership of the mathematics by getting them actively involved.

In this book we have tried to supply you with material for all these ways of using history. The next part, "The History of Mathematics in a Large Nutshell," provides a concise overview of mathematical history from earliest times to the start of the 21st century and establishes a chronological and geographical framework for individual events. The thirty Sketches open up a deeper level of understanding of both the mathematics and the historical context of each topic covered. Finally, "What to Read Next" and the bibliographical remarks sprinkled throughout the book suggest a vast array of resources for you or your students to use in pursuing further information about any of the ideas, people, or events that interest you.

Of course, there's much more to be said about how history may play a role in the mathematics classroom. In fact, this was the subject of a study sponsored by the International Commission on Mathematical Instruction (ICMI). The results of the study were published in [60]. This is not an easy read, but it contains lots of interesting ideas and information. Many recent books combine history with pedagogy; some are collections of articles, such as [169], [23], [98], and [93]. There is also an international society called known as HPM, whose full name is *International Study Group on the Relations Between History and Pedagogy of Mathematics*. The American section, HPM-Americas, runs regular meetings where both history and its use in teaching are explored.

In its journal *Mathematics Teacher*, the NCTM often publishes historical articles that contain ideas about how that history can be used in the classroom. Several packages of classroom-ready modules have been published, including [97] and our own [15] and [16].

The History of Mathematics in a Large Nutshell

The story of mathematics spans several thousand years. It begins as far back as the invention of the alphabet, and new chapters are still being added today. This overview should be thought of as a brief survey of that huge territory. Its intention is to give you a general feel for the lay of the land and perhaps to help you become familiar with the more significant landmarks.

Much (but by no means all) of the mathematics we now learn in school is actually quite old. It belongs to a tradition that began in the Ancient Near East, then developed and grew in Ancient Greece, India, and the medieval Islamic Empire. Later this tradition found a home in late-Medieval and Renaissance Europe, and eventually became mathematics as it is now understood throughout the world. While we do not entirely ignore other traditions (Chinese, for example), they receive less attention because they have had much less direct influence on the mathematics that we now teach.

Our survey spends far more time on ancient mathematics than it does on recent work. In a way, this is a real imbalance. The last few centuries have been times of great progress in mathematics. Much of this newer work, however, deals with topics far beyond the school mathematics curriculum. We have chosen, rather, to pay most attention to the story of those parts of mathematics that we teach and learn in school. Thus, the survey gets thinner as we come closer to the present. On the other hand, many of the topics we might have mentioned there appear in the sketches that make up the rest of this book.

The study of the history of mathematics, like all historical investigation, is based on sources. These are mostly written documents, but sometimes artifacts are also important. When these sources are abundant, we are reasonably confident about the picture of the period in question. When they are scarce, we are much less sure. In addition, mathematicians have been writing about the story of their subject for

5

many centuries. That has sometimes led to "standard stories" about certain events. These stories are mostly true, but sometimes historical research has changed our view of what happened. And sometimes historians are still arguing about the right story. In order to stay short, this survey ignores many of these subtleties. To make up for this, we provide references where you can find more information. To help you on your way, we have also provided an annotated list of books that might be good points of entry for further study. (See "What to Read Next," starting on page 223.)

As you read through this overview, you may be struck by how few women are mentioned. Before the 20th century, most cultures of Western civilization denied women access to significant formal education, particularly in the sciences. Moreover, even when a woman succeeded in learning enough mathematics to make an original contribution to the field, she often had a very hard time getting recognized. Her work sometimes ended up being published anonymously, or by another (male) mathematician who had access to the standard outlets for mathematical publication. Sometimes it wasn't published at all. Only in recent years have historians begun to uncover the full extent of these obscured mathematical achievements of women.[1]

In our times, most of the barriers to women in the sciences have been dissolved. Unfortunately, some of the effects of the old "uneven playing field" still persist. The perception that mathematics is a male domain has been a remarkably resilient self-fulfilling prophecy. But things are changing. The results of careful historical research and the outstanding achievements of many 20th century female mathematicians show that women can be creative mathematicians, have made substantial contributions to mathematics in the past, and will certainly continue to do so in the future.

Beginnings

No one quite knows when and how mathematics began. What we do know is that in every civilization that developed writing we also find evidence for some level of mathematical knowledge. Names for numbers and shapes and the basic ideas about counting and arithmetical operations seem to be part of the common heritage of humanity everywhere. Anthropologists have found many prehistoric artifacts that can, perhaps, be interpreted as mathematical. The oldest such artifacts were

---

[1]Good sources of information are [79], [102], [27], [134], [87], [138], and [149].

found in Africa and date as far back as 37,000 years. They show that men and women have been engaging in mathematical activities for a long time. Modern anthropologists and students of ethnomathematics also observe that many cultures around the world show a deep awareness of form and quantity[2] and can often do quite sophisticated and difficult things that require some mathematical understanding. These range all the way from laying out a rectangular base for a building to devising intricate patterns and designs in weaving, basketry, and other crafts. These mathematical (or pre-mathematical) elements of current pre-literate societies may be our best hint at what the earliest human mathematical activity was like.

By about 5000 B.C., when writing was first developing in the Ancient Near East, mathematics began to emerge as a specific activity.[3] As societies adopted various forms of centralized government, they needed ways of keeping track of what was produced, how much was owed in taxes, and so on. It became important to know the size of fields, the volume of baskets, the number of workers needed for a particular task. Units of measure, which had sprung up in a haphazard way, created many conversion problems that sometimes involved difficult arithmetic. Inheritance laws also created interesting mathematical problems. Dealing with all of these issues was the specialty of the "scribes." These were usually professional civil servants who could write and solve simple mathematical problems. Mathematics as a subject was born in the scribal traditions and the scribal schools.

Most of the evidence we have for this period in the development of mathematics comes from Mesopotamia, the area between the Tigris and Euphrates rivers in what is now Iraq, and from Egypt, the land in the valley of the Nile, in northeast Africa. It is likely that a similar process was happening at about the same time in India and in China, though we have far less evidence about the specifics.

The ancient Egyptians wrote with ink on papyrus, a material that does not easily survive for thousands of years. In addition, most Egyptian archaeological digs have been near stone temples and tombs, rather than at the sites of the ancient cities where mathematical documents are most likely to have been produced. As a result, we have only a few documents that hint at what ancient Egyptian mathematics was like. Thus, our knowledge is sketchy and scholars are not in complete agreement about the nature and extent of Egyptian mathematics. The

---

[2]Good references for this are [8] and [65]. The second includes many ideas for how to use some of this material in the classroom.

[3]For a theory about how this happened, see [153].

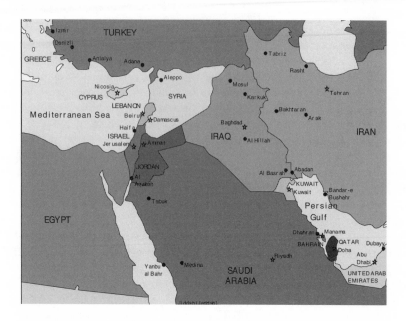

situation with respect to the Mesopotamian cultures is quite different. These people wrote by imprinting clay tablets with a wooden stylus. Many of those tablets have survived, and careful study has led to a much more detailed — though still incomplete and controversial — understanding of their mathematics. These two civilizations existed at about the same time and certainly interacted at several levels, from war to commerce. This contact must have led to some communication of mathematical ideas, but it is hard to trace exactly what was shared. In any case, what has come down to us from these cultures is very different in style and content.

The most extensive source of information about Egyptian mathematics is the Rhind Papyrus, named after A. Henry Rhind, the 19th century archaeologist who brought it to England. It dates back to about 1650 B.C. The papyrus contains, on one side, extensive tables that were used as aids to computation (particularly multiplication) and, on the other side, a collection of problems probably used in the training of scribes. The examples cover a wide range of mathematical ideas but stay close to the sorts of techniques that would be needed by the scribe to fulfill his duties. From this source and others, we can deduce some basic features of ancient Egyptian mathematics:

- The Egyptians used two numeration systems (one mostly for writing on stone, the other for writing on papyrus). Both were based on grouping by tens. One system used different symbols for various powers of ten. Multiples of a particular power were shown by repeating the symbol as many times as needed. For instance, | and ∩ stood for *one* and *ten*, respectively, so 57 was represented by ∩∩∩∩∩|||||||. The method is essentially the same as in Roman numerals, except that only powers of ten are used. The other system was more complicated, still based on powers of ten but with many more symbols. (See Sketch 1.)

- Their basic arithmetic operations were adding and doubling. To multiply or divide, they used an ingenious method based on doubling. (The basic idea is still used today in computer algorithms.)

- Rather than working with fractions, they worked only with the idea of "an nth part." They would speak of "the third" (meaning $1/3$) and "the fourth" (meaning $1/4$). What we would describe as "other fractions" they would express as sums of such parts. For example, what we call "three fifths" they would call "the half and the tenth." Since doubling was so important in their mathematics, one of the numerical tables in the Rhind Papyrus is a table listing the doubles of the parts. For example, the double of the fifth (i.e., $2/5$) is the third and the fifteenth (i.e., $1/3 + 1/15$). Scholars are still arguing about exactly how these expressions were computed. (See Sketches 4 and 9.)

- They could solve simple linear equations. (See Sketch 9.)

- They knew how to compute or approximate the areas and volumes of several geometric shapes, including circles, hemispheres, and cylinders. Perhaps the most difficult geometry problem discussed in the known sources (this time not the Rhind papyrus) is the (correct) computation of the volume of a truncated square pyramid. For some shapes, all they could give were approximations. For example, the area inside a circle was approximated as follows: Take the diameter of the circle, remove "the ninth part" of it, and find the area of the square with the resulting side length. In our terms, this says that the area inside a circle of diameter d is $(\frac{8}{9} \cdot d)^2$, which is actually a pretty good approximation. (See Sketch 7.)

The Rhind Papyrus was used to train young scribes, so it is a bit hazardous to draw conclusions about the whole of Egyptian mathematics from it. Still, we can say that the Egyptian mathematics of 4,000 years ago was already a fairly well-developed body of knowledge with content very similar to some of what we learn about calculation and geometry in elementary and high school today. It was recorded and taught by means of problems that were intended as examples to be imitated. Most of the problems seem to have their roots in the actual work of the scribes. A few, however, seem designed to give young scribes a chance to show their prowess at difficult or complicated computations. It is unclear to what extent the Egyptian mathematicians developed their science beyond what was needed for everyday work, and we also know next to nothing about how their methods were discovered.

The history of ancient Iraq spans thousands of years and a number of cultures, including Sumerian, Babylonian, Assyrian, Persian, and eventually Greek. All of these cultures knew and used mathematics, but there was a lot of variety. Most of our information about the mathematics of Mesopotamia comes from tablets produced between 1900 and 1600 B.C., sometimes called the Old Babylonian period. For this reason, one sometimes refers to the mathematics of this region as Babylonian mathematics. Unlike what happens for Egyptian mathematics, a great many such tablets have been discovered. Once again, most of them seem to be school texts. The abundance of these texts allows us to develop a clearer picture of what Mesopotamian mathematics was like, though of course many mysteries remain.

The mathematical activity of the Babylonian scribes seems to have arisen from the everyday necessities of running a central government. Then, in the context of the scribal schools, people became interested in the subject for its own sake, pushing the problems and techniques beyond what was strictly practical. Like a musician who is not satisfied with playing at weddings and graduations, the well-trained scribe wanted to go beyond everyday problems to something more elaborate and sophisticated. The goal was to be a mathematical virtuoso, able to handle impressive and complex problems. Supporting this was an ideology that saw the scribe as someone who would establish justice and ensure fairness by being able to deal correctly with measurement and quantity. The ability to solve complex problems was a guarantee that the scribe could perform this duty.

Most of the mathematical tablets from this period are either tables to assist in computation or collections of problems for training young scribes. Some of the problem tablets contain answers or full solutions, but there is very little that would explain the discovery process behind the methods being demonstrated. Scholars have developed a good picture of what those methods might have been, but, like all historical reconstructions, the picture involves a large dose of conjecture. Still, there are several things one can say with some certainty:

- In their calculations, the Mesopotamian scribes often represented numbers using a place-value system based on sixty. Repetitions of a ones symbol and a tens symbol were used to denote the numbers 1 through 59. The positions of these groups of symbols relative to each other indicated whether they stood for units or 60s or 60^2s, etc. (See Sketch 1.)

- They made use of extensive tables of products, reciprocals, conversion coefficients, and other constants. Fractions were often expressed in "sexagesimal" format. This is analogous to our way of writing fractions as decimals, but it used powers of 60 instead of powers of 10. (See Sketch 4.)

- Like the Egyptians, the Babylonian scribes could handle linear equations. They could also solve a wide range of problems that we would describe as leading to quadratic equations. Many of these problems are quite artificial and may have existed solely as a way for scribes to demonstrate their prowess. The ideas behind the methods for solving quadratic equations were probably based on a "cut-and-paste geometry" in which pieces of rectangles and squares were moved around to discover the solution. The solutions in the tablets, however, are entirely numerical and are meant to drill students in applying the method. (See Sketch 10.)

- Babylonian geometry, like that of the Egyptians, was devoted mainly to measurement. They appear to have known and applied instances of what we now call the Pythagorean Theorem, and they had formulas for computing or approximating areas and volumes of various common shapes.

One interesting aspect of Babylonian mathematics is the occurrence of problems that do not even attempt to be practical, but instead have a recreational flavor. These are fanciful problems that usually reduce to solving a linear or quadratic equation. Here's an example:

A trapezoidal field. I cut off a reed and used it as a measuring reed. While it was unbroken I went 1 three-score steps along the length. Its 6th part broke off for me, I let follow 72 steps on the length. Again 1/3 of the reed and 1/3 cubit broke off for me; in 3 three-score steps I went through the upper width. I extended the reed with that which [in the second instance] broke off for me, and I made the lower

width in 36 steps. The surface is 1 bur. What is the original
length of the reed?[4]

Except for the rather strange language and the fact that most of us do
not know how many square cubits make up a "bur," this is a problem
that could still appear in many a "recreational math" column — and
it's still quite hard. Puzzles like this one continue to appear throughout
the history of mathematics.

The end of the Old Babylonian period brought many social and
political changes. The ideology of justice through fair measurement
became less influential, and the organization of scribal schools seems to
have changed. It is possible, too, that other kinds of writing material
came into use, so fewer records were preserved. The mathematical work
that survives seems less exciting and less accomplished. Some tablets
mix mathematics with several other subjects. Mathematics loses its
separate identity, and most of the enthusiasm and creativity disappear.
Only much later, around 300 B.C., do we see a resurgence of interest in
mathematics, this time in the service of Babylonian astronomy.

The overall impression is that Babylonian mathematics was driven
by *methods*. Once a method for solving a certain kind of problem was
on hand, the scribes seemed to revel in constructing more and more
elaborate problems that could be solved by that method. Keep in
mind, however, that most of what we have are tablets for training young
scribes; we might get a similar impression from our own textbooks!

Babylonian mathematics has several impressive features, in particular the solution of quadratic equations. Their representation of numbers
in terms of powers of 60 was also very important, particularly in regard
to fractions. The fact that we still divide an hour into 60 minutes and
a minute into 60 seconds goes back, via the Greek astronomers, to the
Babylonian sexagesimal fractions; almost 4000 years later, we are still
influenced by the Babylonian scribes.

Until very recently, we did not know
a lot about very early Chinese mathematics. Before the invention of paper, the Chinese wrote on wood or bamboo strips, which were often tied together with string. These materials are
highly susceptible to decay, so few mathematical texts from before 100 A.D. survived. These natural difficulties were

CHINA

---

[4]This text is a lightly modernized form of a translation given in [91], page 30.

sometimes compounded by human perversity. Just after the Imperial unification of China, at the beginning of the Qin Dynasty (about 220 B.C.), it appears that the Emperor ordered that all books from earlier periods be burned, except for official records and books on medicine, divination, forestry, and agriculture, which were considered "useful."[5]

The situation has changed over the last couple of decades, with various older texts being found by archaeologists. In many cases, the bamboo or wooden strips have survived, but not the string that held them together, so scholars find themselves having to figure out which strips go with which and in what order. For example, in 2007 the Yuelu Academy of Hunan University purchased some 1300 bamboo slips from an antiquities dealer, which seem to contain at least six different books. Among them are 231 slips that make up a book on *Shu* (Numbers), dating back to around 210 B.C.

Many of these newly-discovered texts are still being studied by scholars. They reveal a sophisticated mathematical culture. Besides elementary arithmetic, there are problems involving proportions, some that require the Pythagorean Theorem, and other important ideas. We may expect that there will be further discoveries, and that they will continue to enrich our idea of Chinese mathematics.

The best known mathematical texts from China are the *Ten Mathematical Classics*, books studied by civil servants who were expected to demonstrate the ability to solve mathematical problems before they could get their jobs. Like the texts from Egypt and Mesopotamia, they contain problems and solutions. In China, however, the solutions are often presented with a general recipe for solving that type of problem.

The earliest of the Mathematical Classics is usually known as *The Nine Chapters on the Mathematical Art* (in Chinese, *Jiuzhang Suanshu*). The version that has survived was annotated and supplemented by Liu Hui in 263 A.D. Liu's preface says that the material in the book goes back to the 11th century B.C., but he also says that the actual text was put together around 100 B.C. The later date is almost universally accepted, but scholars differ about the earlier one.

The topics in the *Nine Chapters* are quite varied. The problems are presented in the context of practical situations, but they have already been formalized. Some have a recreational flavor. Several are also known in Western mathematics, sometimes in identical form. This presumably reflects cultural contact along the "Silk Road" trade route connecting China to the Western world. Proportionality was a central idea for these early Chinese mathematicians, both in geometry (e.g.,

---

[5] There is some doubt as to how thoroughly the decree was actually carried out.

similar triangles) and in arithmetic (e.g., solving numerical problems by using proportions). Many geometric problems are analyzed by imagining figures being cut up and moved around; because this often involved removing some pieces and putting in others, the Chinese called it the "out-in" method. Most spectacularly, there is a chapter dedicated to solving systems of linear equations by a method that is essentially the same as the one[6] rediscovered by Gauss in the 19th century. The original *Nine Chapters* contains only problems and solutions, but Liu Hui's commentary often gives justifications for the rules used to solve the problems. These are not formal proofs based on axioms, but they are proofs nonetheless. Chinese proofs, from Liu Hui on, usually had this informal character. Together with the other Mathematical Classics, the *Nine Chapters* played a central role in Chinese mathematics. Many later mathematicians wrote commentaries on it and used it as a jumping-off point for further mathematical work.

The Chinese mathematical tradition that started before 100 B.C. continued to develop and grow for many centuries. The extent of the contact between China and the West is still an open question, but it is clear that at least some ideas migrated along the Silk Road or influenced mathematicians in India. Nevertheless, Chinese mathematics remained quite independent (and different) until European explorers arrived in the 16th century. Because the influence of Chinese ideas on Western mathematics was so indirect, we will not discuss it further in detail.

We know even less about very early Indian mathematics. There is evidence of a workable number system used for astronomical and other calculations and of a practical interest in elementary geometry. The most significant early texts are the *Vedas*, a large collection of verses that probably achieved its final form around 600 B.C. They come with supplementary texts called the *Śulbasūtras*, which mostly focus on the rules for building altars. That requires some mathematics; we find a statement of the Pythagorean Theorem (see Sketch 12), methods for approximating the length of the diagonal of a square, and lots of discussion about the surface areas and volumes of solids. Other early sources show an interest in very large numbers and hint at other mathematical developments that almost certainly played a role in later developments in India. The Indian tradition influenced Western mathematics quite directly, so we will come back to it later.

---

[6]It is known as "Gaussian elimination" or "row reduction."

Were there contacts between these civilizations, and did the mathematics of one influence the other(s)? In many cases, there certainly was contact of some form, but it is hard to tell whether mathematical ideas were transmitted. Interest in non-Western mathematics has grown[7] in recent years, but a scholarly consensus on the spread and transmission of ideas has not yet developed.

There are good introductory articles on ancient Egyptian, Mesopotamian, Chinese, and Indian mathematics in Part 1 of [74] and in [157]. For a readable, full-length survey, look at [94], which contains both an account of non-Western mathematics and a passionate argument for its influence and importance. Robson's [147] is a detailed account of how mathematics fit into the social structures of Ancient Mesopotamia. Plofker's [140] surveys the whole tradition of mathematics in India, from the earliest stages to the late Medieval period. For extensive information about Chinese mathematics, see [120] or [115]. Papers on the notion of proof in various ancient cultures are collected in [28]. Finally, nothing replaces reading the real thing: Selected, translated, and annotated mathematical texts from non-Western cultures can be found in [96].

Greek Mathematics

Many ancient cultures developed various kinds of mathematics, but the Greek mathematicians were unique in putting logical reasoning and proof at the center of the subject. By doing so, they changed forever what it means to do mathematics.

We do not know exactly when the Greeks began to think about mathematics. Their own histories say that the earliest mathematical arguments go back to 600 B.C. The Greek mathematical tradition remained a living and growing endeavor until about 400 A.D. There was much change and growth during those thousand years, of course, and historians have worked hard to understand the process that led to the particular Greek slant on the subject. This task is made harder by the fact that most of our sources of information are rather late. Apart from some remarks in Plato and Aristotle and some fragments, our earliest witness to Greek mathematics is Euclid's *Elements*, which dates from around 300 B.C. Much of our information on the history of Greek mathematics is even more recent, coming from the 3rd and 4th centuries A.D. These texts probably preserve earlier material, but it's hard to be

---

[7]For example, see the proceedings of a conference on "2000 Years of Transmission of Mathematical Ideas" in [47].

sure. A lot of scholarly detective work has gone into reconstructing the overall history, and the issues are still far from settled. Our account here can only scratch the surface of the vast body of scholarship on Greek mathematics.

It is important to stress that when one speaks of "Greek mathematics" the main reference of the word "Greek" is to the *language* in which it is written. Greek was one of the common languages of much of the Mediterranean world. It was the language of commerce and culture, spoken by all educated people. Similarly, the Greek mathematical tradition was the dominant form of theoretical mathematics. It is certain that not all "Greek" mathematicians were born in Greece. For example, Archimedes was from Syracuse (in Sicily, now a part of Italy), and Euclid is traditionally located in Alexandria (in Egypt). In most cases, we know nothing about the actual ethnicity, nationality, or creed of these mathematicians. What they had in common was a tradition, a way of thought, a language, and a culture.

Like most Greek philosophers, the mathematicians of the earliest period seem to have been people of independent means who spent their time on scholarly pursuits. Later, some mathematicians made a living as astrologers, a few were supported by the state in one way or the other, and some seem to have done some teaching (usually one-on-one, rather than in schools). On the whole, however, mathematics was a pursuit for those who had the means and the time — and, of course, the talent. The total number of working and creative mathematicians at any given time was probably very small, maybe a dozen or so.[8] Mathematicians worked mostly alone and communicated with each other in writing. Despite this, they built an intellectual tradition that continues to impress everyone who comes in contact with it.

The dominant form of Greek mathematics was geometry, though the Greeks also studied the properties of whole numbers, the theory of ratios, astronomy, and mechanics. The latter two were treated very much in geometric and theoretical style. There was no sharp dividing line between "pure" and "applied" mathematics. (In fact, that distinction dates back only to the 19th century.) Most Greek mathematicians had little interest in practical arithmetic or in the problems of actually measuring lengths and areas. These issues only came to the fore relatively late (for example, during the 1st century A.D. in the work of Heron, who may have been influenced by Babylonian mathematics),

---

[8]Reviel Netz estimates that the total number of mathematicians during the whole thousand-year period of Greek mathematics was no more than 1,000. Of those, maybe 300 were known by name late in the Greek period; about 150 are mentioned in the surviving texts. See [131, chapter 7].

and they remained to some extent a separate tradition.

According to the ancient Greek historians of geometry, the first Greek mathematicians were Thales, who lived around 600 B.C., and Pythagoras, a century later. When those histories were written, both Thales and Pythagoras were already mythic figures from the distant past. Many stories are told about them, and it is hard to know which, if any, of these stories contain any historical truth. Both men are said to have learned their mathematics in Egypt and in Mesopotamia. Thales is said to have been the first person to attempt to prove some geometrical theorems, including the statements that the sum of the angles in any triangle is equal to two right angles, the sides of similar triangles are proportional, and a circle is bisected by any of its diameters.

Later Greek authors told many stories about Pythagoras. The legends center on a semi-religious society called the Pythagorean Brotherhood (even though women were virtually equal members). The home base of the Pythagoreans was probably Crotona, a city founded by Greek settlers in southern Italy. The Brotherhood was a secret society dedicated to learning of various kinds, mostly religious and philosophical.

Lots of strange stories were passed on about the Brotherhood. Most of them were written down centuries after the fact by "neo-Pythagorean" philosophers. They describe a society some of whose customs would strike us today as strange, others as eminently sensible. The members apparently never ate meat or beans, never hunted, used no wool, dressed in white, and slept in white linen bedding. They had a variety of rituals to strengthen their sense of community, and the pentagram was their symbol. They believed in a sort of reincar-

pentagram

nation and developed a kind of number mysticism, a belief that numbers were the secret principle of reality. Each day followed a common, simple regimen designed to strengthen both mind and body. They exercised to keep physically fit, had periods of silent contemplation, and spent a large part of each day discussing and studying *mathematike*, "that which is learned."

Many of the ideas and achievements of the later Pythagoreans probably were eventually attributed to Pythagoras himself. Most scholars believe that Pythagoras himself was not an active mathematician, though he may have been interested in number mysticism. But at some later point it seems that some Pythagoreans began to construct formal arguments, and hence to do mathematics. Because the Pythagoreans remained influential for some time, we know (or can guess at) some of their mathematical ideas. They seem to have been much concerned with the properties of whole numbers and the study of ratios (which they related to music). In geometry, they are, of course, credited with the Pythagorean Theorem. (See Sketch 12.) It is likely, however, that the most important success often credited to the Pythagoreans is the discovery of *incommensurable ratios.*

Ratios played a very important role in Greek mathematics, because the Greek geometers did not directly attach numbers to the objects they studied. A line segment was a line segment. There are equal line segments, longer and shorter line segments, and a segment might be equal to two others put together, but at no time did the Greek mathematicians talk about the *length* of a line segment.[9] Areas, volumes, and angles were treated as different kinds of quantities, none of which was necessarily connected to any numbers. So how does one compare quantities? What the Greek mathematicians did was to work with *ratios* of quantities. For example, to find the area of a circle, we use a formula, $A = \pi r^2$, which tells us to take the *length* of the radius and multiply it by itself and then by a constant we call π. The result will be a number, which we call the area of the circle. The Greeks expressed the same idea by saying:

> The ratio between two circles is the same as the ratio between two squares with sides equal to the radii of the circles.

In our language we would say "the areas of two circles" and "the areas of two squares." We would probably also use symbols: If A_1 and A_2 are the areas of the two circles and r_1 and r_2 are the two radii, then

$$\frac{A_1}{A_2} = \frac{r_1^2}{r_2^2}.$$

It follows that

$$\frac{A_1}{r_1^2} = \frac{A_2}{r_2^2}.$$

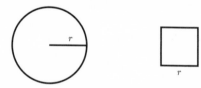

A circle of radius r and a square with side r

That is, the ratio of a circle to the square whose side is equal to the radius (i.e., A/r^2) is always the same, regardless of the size of the circle. We now regard this ratio as a number, which we call π, and we know that it is quite a complicated number. (See Sketches 7 and 29.)

---

[9]Of course, in everyday life lengths were computed and measured in Greece as they were everywhere else. In the Greek tradition there was a real gap between theoretical mathematics and everyday use of mathematical ideas.

Some ratios are easy to understand because they are equal to the ratios of two whole numbers. If a segment is equal to twice another segment, their ratio is 2 to 1. Similarly, it is easy to understand what is meant by two segments whose ratio is 3 to 2. The great insight of the Pythagoreans was to see that the ratio of two segments will not always be that simple. In fact, they proved that the ratio between the side and the diagonal of a square cannot be a ratio of *any* two whole numbers. They called segments of this kind *incommensurable*, and they called the ratios between such segments *irrational*.[10] (See Sketch 29 for more on incommensurability and irrational numbers.)

By the time of the philosophers Plato and Aristotle, knowing about the existence of incommensurable segments was part of every cultured person's education. Both philosophers used it as an example of something that was not evident to the senses but could be discovered by reason. Both also showed great interest in and respect for mathematics. It is said that Plato had a sign at the door of his Academy saying, "Let no one ignorant of geometry enter here." The story may not be true, since the earliest surviving text that mentions the inscription was written more than 700 years after Plato's time. Nevertheless, such an inscription would be consistent with his attitude towards mathematics. For example, he listed mathematics as a fundamental part of the ideal education in his *Republic*. He also mentions mathematical results in several of his dialogues. (See Sketch 15 for a description of the "Platonic Solids.")

Aristotle also mentions mathematics frequently in his works. For instance, he uses mathematical examples when discussing correct reasoning. This suggests that by his time mathematicians were already engaged in working out formal proofs of mathematical statements. Around this time, they probably began to understand that in order to prove theorems one must start with a few unproved assumptions. In fact, Aristotle says so explicitly. These basic assumptions, or postulates, were taken for granted and accepted as true. This is the structure of the oldest Greek mathematical work that still survives (mostly) intact, the *Elements* of Euclid.

Between the time of Plato and Aristotle and the time of Euclid, an important change in Greek culture had occurred. The two philosophers, and probably most of the mathematicians associated with them, lived in Athens, one of the centers of Ancient Greek civilization. In

---

[10]Well, they actually used the Greek words *alogos* and *arrhetos*, which could also mean "unspeakable" or "inexpressible." But "irrational," meaning "without ratio," is the word that prevailed historically.

Aristotle's time, however, Alexander the Great had set out to conquer other peoples and create a great empire. In so doing, he spread the Greek language, culture, and learning to many other parts of the world. For several centuries to follow, most educated people living in the countries bordering the Eastern Mediterranean spoke Greek. In effect, Greek became the international language of that region. Greek learning also spread. It flourished in a spectacular way in northern Egypt, in a city called Alexandria (one of many cities named for Alexander the Great), near the mouth of the Nile. The mathematical tradition there was particularly strong. Perhaps it was connected to the famous temple to the Muses (the *Museum*) and to its equally famous library. By the end of the 4th century B.C., Alexandria was the real center of Greek mathematics.

Beyond the fact that he probably lived in Alexandria around 300 B.C., we know almost nothing about Euclid himself. What we have are his writings, of which the most famous is a book called *Elements*. It is a collection of the most important mathematical results of the Greek tradition, organized systematically and presented as a formal deductive science. The style is dry and efficient. The book opens with a list of definitions, followed by postulates and "common notions" (which Euclid takes as self-evident). After that comes a sequence of propositions, each followed by a proof. There is no connecting material, no attempt at motivation. Next to each proposition is a diagram, and the proof typically refers to it in a crucial way. (See Sketch 14.) Later, as he moves on to other subjects, Euclid introduces more definitions and postulates. In this way, he covers plane and solid geometry, studies the divisibility properties of whole numbers, works out a sophisticated theory of ratios (twice, in fact; once for magnitudes and once for whole numbers), and develops a complicated classification of quadratic irrational ratios. The *Elements* brings together in one place the main accomplishments of Greek mathematics up to that time.

The *Elements* was a massive achievement, and its style and content were enormously influential, not only on Greek mathematics, but also on the Western mathematical tradition. Studying the first portion of the *Elements* became an intellectual rite of passage in the West, even as recently as the early part of the 20th century. Euclid's book was held up as a model of clear and precise reasoning, and was imitated by others who aspired to rigor and precision.

Greek geometry does not end with the *Elements*. Euclid himself also wrote books on conic sections, geometric optics, spherical geometry, and solving geometrical problems. Archimedes wrote about areas and volumes of various curved figures, and Apollonius wrote a treatise on conic sections that is still an impressive display of geometric prowess. (See Sketch 28.) Geometry continued to be a central interest of Greek mathematicians for several centuries.

The systematic and ordered presentation of mathematical results that we see in the *Elements* is only one part of the Greek tradition. Another important component (some claim it's the most important) is the tradition of mathematical problems. In fact, at times one can see such problems "sitting behind" Euclid's text. For example, since the Greeks did not measure areas by assigning numbers to them, they attacked the measurement of areas by trying to construct a rectangle (or square) whose area is the same as the area of a given figure. For figures bounded by straight lines, this is done in the first two sections (known as "books") of Euclid's *Elements*. From this point of view, the Pythagorean Theorem can be seen as a way to construct a square equal to the sum of two other squares. By the end of Book II, he has shown that he can construct a square equal to any given polygonal figure.

What happens, however, when one tries to do the same for a circle? That is, given a circle, can one construct a square with the same area as the circle? This is called the problem of *squaring the circle*, and it turns out to be very hard. In fact, it leads quickly to the question of what we mean by "construct." Greek mathematicians knew how to solve the problem of squaring the circle provided they were given a specially defined curve (either a *quadratrix* or an *Archimedean spiral*). But is this really a solution to the problem? Some mathematicians objected that constructing these curves was itself problematic, so the problem had not really been solved.

Besides the problem of squaring the circle, two other problems became famous in Greek times. One was *trisecting the angle* — constructing an angle that is one-third of a given angle. The other was *duplicating the cube* — constructing a cube whose volume is twice the volume of a given cube. Greek mathematicians eventually solved both problems. Their solutions, however, always involved using some sort of auxiliary device, sometimes a mechanical device, sometimes a mathematical one. (See Sketch 28 for how the problem of duplicating the cube is related to conic sections.) Later mathematicians reinterpreted the problems by adding the requirement that the constructions use only a ruler and a compass, i.e., using only lines and circles. We now know that under this constraint neither construction is possible, but this was not proved until the 19th century. Some Greek mathematicians knew (or suspected) this too, though they could not prove it. For example, Pappus (writing ca. 320 A.D.) criticizes a proposed ruler and compass solution of the problem of duplicating the cube by saying that everyone knew that this problem required other techniques.

These and other problems were among the main motivations behind Greek geometry. Theorems seem often to have been discovered as steps towards solving some problem. In fact, some Greek geometric problems were so difficult that they served as motivation for the development of coordinate geometry by the mathematicians of the 17th century.

 While geometry was the central topic in Greek mathematics, many other subjects also show up. There was also a lot of interest in astronomy, and an elaborate *spherical geometry* (the geometry of the surface of a sphere) was developed to explain and predict the movement of the stars and planets. To be able to locate a planet in the sky one needs to have a way of measuring angles, so numbers and magnitudes cannot be kept completely separate. Greek astronomers borrowed from Babylonian mathematics and started using numbers to measure angles. One can see the Babylonian connection because fractions of angles were written in sexagesimal fashion, as they still are today, in "minutes" and "seconds" of angle. Also in this context we see the beginnings of trigonometry. (See Sketches 18 and 26.)

The most famous Greek astronomer was Claudius Ptolemy, who lived in Alexandria around 120 A.D. He wrote on many subjects, from astronomy and geography to astrology, but his most famous work is the *Syntaxis*, known today by the nickname given to it by Arabic scholars many centuries later. They called it *Almagest*, derived from a Greek term meaning "the greatest." Ptolemy's book is an amazing achievement; it provides a workable and accurate description of all visual astronomical phenomena. It was the basis of almost all positional astronomy until the 16th century.

Diophantus, who probably lived a century or so after Ptolemy, was one of the most original of the Greek mathematicians. His *Arithmetica* contains no geometry and no diagrams, focusing instead on solving numerical problems. The book is simply a list of problems and solutions. In the problems, Diophantus used a notation for the unknown and its powers that hints at the algebraic notation developed a thousand years later in Europe. His problems always asked for numbers, which to him meant rational numbers (common fractions). For example, one problem asks for a way to write a square as the sum of two other squares. In his solutions, Diophantus always used specific numbers and proceeded to explain how to find a solution. When solving the problem of writing a square as the sum of two squares, he started by saying "suppose the square is 16." Then he went through several steps and ended up with

$$16 = \frac{256}{25} + \frac{144}{25} = \left(\frac{16}{5}\right)^2 + \left(\frac{12}{5}\right)^2.$$

While his solutions use specific numbers, they are intended to be general. The reader is supposed to see that a similar process works for any initial number chosen. One interesting feature of this approach is that, every once in a while, there will be a problem that is solvable for some initial numbers but not for others. In such cases, Diophantus usually worked out the conditions under which his problems are solvable, thereby confirming that he was trying to find general solutions.

Diophantus's work seems to have been lost and rediscovered many times. Eventually, it had a deep influence on the European algebraists of the 16th and 17th centuries. We still refer to equations that are to be solved in whole numbers or rational numbers as *Diophantine equations*. It's not clear, however, whether it had any impact in his own time.

After about 300 A.D., Greek mathematics lost some of its creative flair. In this period, there begins to be an emphasis on producing editions of and commentaries on older works. These books are actually our best sources for the Greek mathematical tradition, since they collect so much earlier material. At the same time, they create the difficult problem of distinguishing between their additions and comments and the original texts.

Probably the most important of the later Greek mathematicians is Pappus, from the mid-4th century. His *Collection* is a kind of "collected works" that includes original work, commentaries on earlier work, and summaries of the works of other mathematicians. Perhaps the most important part of Pappus's work, from a historical point of view, was his discussion of "the method of analysis." Roughly speaking, "analysis" was the method for discovering a proof or a solution, while "synthesis" was the deductive argument that gave the proof or the construction. Euclid's *Elements*, for example, is all synthesis. Pappus's discussion of analysis is not very specific. This vagueness ended up being important, because the mathematicians of the Renaissance understood him to mean that there was a secret method behind much of Greek mathematics. Their attempts to figure out that method led to many new ideas and discoveries in the 16th and 17th centuries.

After Pappus, most of the significant Greek mathematicians were involved in writing commentaries on earlier work. Theon, who lived in Alexandria in the 4th century, prepared new editions of Euclid's *Elements* and Ptolemy's *Syntaxis*. Theon's daughter Hypatia[11] wrote commentaries on her father's work, on Apollonius's *Conics* and on Diophantus's *Arithmetic*. Hypatia was also famous as a teacher of Platonic philosophy in Alexandria, where Christianity had become the dominant religion. Unfortunately, she became entangled in a power struggle between the Prefect Orestes and the Archbishop Cyril and was brutally murdered by followers of the Archbishop.

Proclus was one of the last important writers in the Greek tradition. He wrote a commentary on parts of Euclid's *Elements* that is heavily influenced by Neoplatonic philosophy. His commentary includes a history of early Greek mathematics, which most scholars feel incorporates portions of a much earlier work by Eudemus.

The 5th century A.D. marks the end of the Greek mathematical tradition in its classical form. Before leaving that period, however, we should note that this Greek tradition was not the only kind of mathematics going on in Greek, Hellenistic, and Roman culture between 600 B.C. and 400 A.D. Beneath the surface, so to speak, of the "scientific" tradition of the mathematicians there was a "subscientific" tradition. This was the mathematics of everyday life. Regardless of the mathematicians' fascination with geometry and disdain for numbers, merchants had to add and subtract, tax collectors had to measure areas of fields, and architects and engineers had to make sure their buildings and bridges didn't fall down. All of this required mathematical knowledge, which seems to have been passed on almost independently of what the mathematical scholars were doing. In fact, much of it was never written down. (The exception is the work of Heron, who sits somewhere in the middle between the scientific and subscientific traditions and tries to get them talking to each other.) One of the interesting features of the subscientific tradition is the presence, once again, of recreational problems, much like the ones in the Babylonian texts.

A vast amount has been written about Greek mathematics. Because it was so influential, it gets a large chunk of every survey of the history of mathematics, and reading the Greek chapters in such a survey (e.g., [99] or [30]) is a good way to start learning more. There is also a brief summary in [74]. The most accessible and most up-to-date book-length survey of Greek mathematics is [35]. Many of the original Greek texts are available in English translation, largely thanks to Thomas L. Heath

---

[11] Hypatia is not the only woman to have been an active mathematician in Greek times. At least one other is known: Pandrosian, a teacher of mathematics to whom the third book of Pappus's *Collection* is addressed.

and Dover Books; for example, see [49] and [50]. One can often find
these texts online. The subscientific tradition is discussed in [91]; for
the recreational aspect, see the work of David Singmaster (e.g., [163]
and [162]). Finally, there are several recent books that have changed
our understanding of Greek mathematics. Typically these are difficult,
but they are often exciting to read; three important examples are [108],
[64] and [131].

Meanwhile, in India

For the next four hundred years or so, Europe and North Africa saw very little mathematical activity. In Western Europe, North
Africa, and the Middle East, barbarian in-
vasions fragmented the Roman Empire. The
resulting social conditions did not favor intellectual activity. The Eastern Roman Empire was still strong and very much under the

influence of Greek culture, but the scholars of Byzantium were more
interested in other things. They preserved copies of the ancient mathematical manuscripts, but only occasionally did anyone show any real
interest in their contents. The advent of Islam in the 7th century further unsettled these areas. In the 8th century, Islamic forces attacked
and conquered all of North Africa, most of the Middle East, and even
parts of Europe. It was only after the Islamic Empire began to settle
down politically that the right conditions for mathematical research
could be found.

Of course, throughout this period people were still building, buying,
selling, taxing, and surveying, so the subscientific tradition certainly
persisted in all of these areas. In many cases, we have little textual
evidence of how mathematics was understood and passed on. The
texts we do have suggest a mathematics that was practical but not
deep. The subject still kept its prestige. Western philosophers would
say nice things about the importance and significance of geometry,
for example. But when they actually explained what geometry they
meant, they presented a mixed bag containing some Greek geometry,
some traditional material on measuring and surveying, and quite a bit
of metaphysical speculation.

During this quiet period in Europe and North Africa, the mathematical tradition of India grew and flourished. As mentioned above,
at the time Greek mathematics was beginning to develop, there was
already a local mathematical tradition in India. It is likely that this

tradition received some influence from earlier astronomers, Babylonian and Greek. Astronomy was, in fact, one of the main reasons for the study of mathematics in India. Many of the problems studied by the Indian mathematicians were inspired by astronomical questions. Having started there, however, the Indians became interested in mathematics for its own sake. Even a playful element comes in.

The earliest surviving mathematical texts from India are written in Sanskrit verse. This required a great many tricky ways of expressing mathematical ideas. Each key term (say, the name of a number or of an operation) could be expressed in several different ways, allowing the poet to choose the one that fit his verse. These texts are therefore very hard to understand! Soon a tradition of writing commentaries explaining older texts was in place, as well.

As in the case of Greek mathematics, there is only a handful of mathematicians whose names we know and whose texts we can study. The earliest of these is Āryabhaṭa, who did his mathematical work early in the 6th century A.D. In the 7th century, the most important mathematicians are Brahmagupta and Bhāskara, who were among the first people to recognize and work with negative quantities. (See Sketch 5.) Probably the most important mathematician of medieval India was another Bhāskara, who lived in the 12th century. (To distinguish the two, most historians talk about Bhāskara I and Bhāskara II.) In almost all cases, the mathematical texts we have are portions of more extended books on astronomy.

The most famous invention of the Indian mathematicians is their decimal numeration system. (See Sketch 1.) From an earlier system, they retained nine symbols for the numbers from 1 to 9. They introduced place value and created a symbol, a dot or a small circle, to denote an empty place. (See Sketch 3.) The result was the numeration system that we still use today. The history of this momentous step is obscure. It seems likely that there was some influence from China, where a decimal counting board was used. In any case, before the year 600 Indian mathematicians were using a place-value system based on powers of ten. They had also developed methods for doing arithmetic with such numbers. All of the mathematicians mentioned above dedicated a portion of their works to explaining the decimal system and giving rules for computation.

The convenience of the new system seems to have been a powerful argument in its favor. It quickly spread to other countries. A manuscript written in Syria in 662 mentions this new method of calculation. There is also evidence that the system was used in Cambodia and in China soon after. By the 9th century, the new numeration system was known in Baghdad, and from there it was transmitted to Europe.

The Indians also made an important contribution to trigonometry. Greek astronomers had invented trigonometry to help them describe the motion of planets and stars. The Indian astronomers probably learned this theory from Hipparchus, a predecessor of Ptolemy. Greek trigonometry revolved around the notion of the *chord* of an angle. To a central angle β in a circle, one attached the line segment determined by the intersection of the sides of the angle with the circle, as in (a) of the display below. This was called the chord of β. It turns out, however, that in many cases the right segment to consider is not the chord, but rather half the chord. The Indian mathematicians used half the chord of twice the angle as their basic trigonometric segment. They called it a "half-chord." This name was mistranslated into Latin (via the Arabic) as "sinus," giving rise to our modern *sine* of α. Indeed, (b) of the display shows that

$$\sin(\alpha) = \tfrac{1}{2}\mathrm{chord}(2\alpha).$$

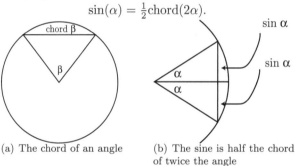

(a) The chord of an angle   (b) The sine is half the chord of twice the angle

The move from chords to sines made trigonometry much simpler than it had been before. (See Sketch 18 for more details.)

However, the sine was thought of as a line segment in a particular circle, not as an abstract number or ratio. Āryabhaṭa, for example, worked with a circle of radius 3438 to construct a table of sines. (Why 3438? Because he wanted the circumference of the circle to be as close as possible to $21600 = 360 \times 60$. That would make one unit of length on the circumference correspond to one minute of arc.) For each angle, the table of sines gave the length of the sine of that angle, that is, of a certain line segment related to that circle. Thus, the sine of 90° was given as 3438. To apply the table to a circle with a different radius, one had to use proportionality to adjust the values.

Since it is almost always impossible to determine the sine exactly, constructing a table of sines requires approximation techniques. Ptolemy had already used such techniques to construct his table of chords.

The Indian mathematicians took his ideas much further, and approximation methods became an important part of their mathematics. Starting from simple ideas, the mathematicians of India eventually developed quite sophisticated formulas for approximate computation.

Indian mathematicians were also interested in algebra and in some aspects of combinatorics. They had methods for computing square and cube roots. They knew how to compute the sum of an arithmetic progression. They handled quadratic equations using essentially the same formula we use today, except that they expressed it in words, usually in verse. Their examples are often problems posed in a playful manner. Here's an example from Bhāskara II:

> The eighth part of a troop of monkeys, squared, was skipping in a grove and delighted with their sport. Twelve remaining monkeys were seen on the hill, amused with chattering to each other. How many were there in all?[12]

In addition to equations in one variable, the Indian mathematicians studied equations in several variables. Such equations usually have many solutions, so to make the problem interesting one must add some conditions, usually specifying that the solutions be of a particular type. A typical example would be to look for solutions that involved only whole numbers. Āryabhaṭa and Brahmagupta could solve linear equations of the form[13] $ax + by = c$, where a, b, and c are whole numbers and the solutions sought are those in which x and y are also whole numbers. Brahmagupta also studied some much harder problems of this kind, such as finding integers x and y for which $92x^2 + 1 = y^2$. Later, Bhāskara II generalized Brahmagupta's ideas. He described a method that will find a solution of $nx^2 + b = y^2$ in whole numbers whenever such a solution exists. Problems of this type are difficult, and the Indian achievement in this area is quite impressive.

Between the 6th and 12th centuries, Indian mathematicians produced a wide range of interesting mathematics. From our perspective, the main thing that is mostly missing from their texts is any explanation of how their methods and results were found. They did not give proofs or derivations. One cannot simply guess at such results, however, so they must have derived them somehow. One conjecture is that they considered the derivations to be trade secrets and so did not write them down.

---

[12] From [99]; the translation is by H. T. Colebrooke.

[13] We give these in modern algebraic form. This is *not* how they were expressed at the time! See Sketch 8.

The Indian tradition did not stop in the 12th century, of course. For example, the Kerala school made amazing discoveries between the 14th and 16th centuries. The later work, however, does not appear to have influenced European mathematics, so we will not discuss it in detail.

India is very far from Europe. There was little contact between European and Indian scholars until the 16th century, and so Western mathematics did not directly learn from the mathematicians of India. Instead, as we discuss in the next section, many of the results found in India made their way to the West through Baghdad and the Arabic mathematical tradition.

The best survey of the history of mathematics in India is [140]. Also interesting is [94], which is particularly interested in the value and influence of non-Western traditions. Both books include a discussion of the later work as well. For shorter surveys, go to [74, Section 1.12], [157], and [99, Chapter 6]. Translations of some of the main sources can be found in [96]. For the brave, there is [101], which gives an annotated translation of our earliest source, Brahmagupta's commentary on the mathematical work of Āryabhaṭa.

Arabic Mathematics

In 750 A.D., the Islamic Empire stretched all the way from the western edge of India to parts of Spain. The period of expansion was coming to an end. A new dynasty, the Abbasids, had just come into power. One of their first actions was to establish a new imperial capital. This new city, called Baghdad, quickly became the cultural center of the Empire. Its location, on the Tigris River in what is now central Iraq, made it a natural crossroads, the place where East and West could meet.

The first scientific works brought to Baghdad were books on astronomy, probably from India. Early in the 9th century, however, the Abbasid caliphs took a more deliberate approach to the intellectual growth of the empire. They obtained scholarly manuscripts in Greek and Sanskrit and brought to Baghdad scholars who could read, understand, and translate them. Over the following years, many important Greek and Indian mathematical books were translated and studied. As a result, a new era of scientific and mathematical creativity began. One of the first Greek texts to be translated was, of course, Euclid's

Elements. It had a huge impact. Once they had learned and absorbed the Euclidean approach, the mathematicians writing in Arabic adopted it wholeheartedly. From then on, many of them formulated theorems precisely and proved them in Euclid's style.

Like the Greek, the Arabic mathematical tradition is marked by the use of a common language. Arabic was the language of scholars in the far-flung empire. Not all of the great mathematicians writing in Arabic were ethnically Arab, and not all of them were Muslim. Their common language allowed them to build on each other's work, creating a new and vital mathematical tradition that was active from the 9th to the 14th centuries.

Muḥammad ibn Mūsa al-Khwārizmī was one of the earliest Arabic mathematicians to make an enduring name for himself. His name indicates that he was from Khwārizm, a town (currently called Khiva) south of the Aral Sea in what is now Uzbekistan. Al-Khwārizmī was active in the middle of the 9th century, and he wrote several books that were to be enormously influential. One was an explanation of the decimal place value system for writing numbers and doing arithmetic, which he said came from India. Three hundred years later, after being translated into Latin, this book was the major source for Europeans who wanted to learn the new numeration system. (See Sketch 1.)

Also by al-Khwārizmī was the book of "al-jabr w'al muqābala," which means something like "restoration and compensation." This book starts off with a discussion of quadratic equations, then goes on to practical geometry, simple linear equations, and a long discussion of how to apply mathematics to solve inheritance problems. It is the portion on quadratic equations that became famous. (See Sketch 10.) Al-Khwārizmī explains how to solve such equations and justifies the method using a naïve geometric approach reminiscent of Mesopotamian mathematics — which is not surprising, since Baghdad is only some 50 miles away from ancient Babylon. When this book was later translated into Latin, "al-jabr" became "algebra."

In these and other books, al-Khwārizmī seems to have been passing on material he had learned from various sources. He had learned decimal numeration from India. The sources of his algebra book are less clear. There was some influence from India, some from Hebrew mathematics, and probably some from the native Mesopotamian tradition. A good bit of the material may have been part of the subscientific, practical tradition. If so, al-Khwārizmī's work is an instance of scientific mathematics learning from the subscientific tradition.

After al-Khwārizmī, algebra became an important part of Arabic mathematics. Some mathematicians worked on the foundations of the subject, giving Euclidean-style proofs that algebraic methods worked. Others extended the methods. Arabic mathematicians learned to manipulate polynomials, to solve certain algebraic equations, and much more. All of this was done using no symbols at all. Arabic algebra was done entirely in words. For example, to express an equation such as $3x^2 = 4x + 2$, they would say something like "three properties[14] are equal to four things plus two dirhems." The solution would similarly be spelled out in words. (See Sketch 10 for more details.)

One of the most famous Arabic mathematicians was 'Umar al-Khāyammī, known in the West as Omar Khayyám. Nowadays mostly known as a poet, in his day he was also famous as a mathematician, scientist, and philosopher.[15] Khayyám also wrote a book on algebra. One of his goals was to find a way to solve equations of degree 3. He was unable to find a numerical solution, but he did find a way to solve all such equations using geometric constructions. (Greek mathematicians had done this for some cubic equations, as we discuss in Sketch 28, but Khayyám does it systematically.) His solution was obtained by going beyond ruler and compass, using parabolas and hyperbolas to determine points. He notes in his book, however, that if one wants to find a number that solves the equation, this geometric solution isn't very helpful. The challenge he laid down with these words was to be taken up by Italian algebraists many centuries later. (See Sketch 11.)

To the Arabic mathematicians, only positive numbers made sense. On the other hand, they were much more willing than the Greeks to consider lengths of various line segments as numbers. In part, this is due to their interest in trigonometry: tables require numbers. Trigonometry also led them to notice that by choosing a fixed segment as the unit one could obtain the lengths of other segments in terms of ratios.

The Arabs did significant work in geometry and trigonometry. They investigated the foundations of geometry, focusing in particular on Euclid's fifth postulate. (See Sketch 19.) They also undertook original geometrical investigations, extending the work of the Greeks. Trigonometry was a major concern, mostly because of its applications to astronomy. The work on trigonometry led inevitably to work on approximate solutions of equations. A particularly notable instance of this is

---

[14]Squares were "properties" because they were areas.

[15]A few scholars argue that al-Khāyammī the poet and al-Khāyammī the philosopher/mathematician were actually two different people, but this is a minority position.

a method for approximating the nth root of a number, developed by al-Kashi in the 14th century.

Combinatorics also shows up in the Arabic tradition. They knew at least the first few rows of what we now call "Pascal's Triangle," and they understood both the connection with $(a+b)^n$ and the combinatorial interpretation of these numbers. Stimulated by their translations of Euclid and Diophantus, they also did work in number theory. There is hardly any part of mathematics on which they did not make their mark.

$$
\begin{array}{ccccccccccccc}
 & & & & & & 1 & & & & & & \\
 & & & & & 1 & & 1 & & & & & \\
 & & & & 1 & & 2 & & 1 & & & & \\
 & & & 1 & & 3 & & 3 & & 1 & & & \\
 & & 1 & & 4 & & 6 & & 4 & & 1 & & \\
 & 1 & & 5 & & 10 & & 10 & & 5 & & 1 & \\
1 & & 6 & & 15 & & 20 & & 15 & & 6 & & 1 \\
 & & & & & & \vdots & & & & & &
\end{array}
$$

Pascal's Triangle: Each inside number is the sum of the two numbers diagonally above it.

Finally, it is important to mention that practical mathematics was also advancing. One example may be due, at least in part, to the Islamic prohibition against graven images, which included any artistic representation of the human body. As a result, a complex and sophisticated art of ornamentation was developed. Buildings were decorated by repetitions of a simple basic motif. This kind of decoration requires some level of forethought, because not all shapes can be repeated in such a way as to cover a plane surface. Deciding what sorts of shapes can be used in this way is really a mathematical question, linked both to the study of plane tilings and the mathematical theory of symmetry. There is no evidence, however, that Arabic mathematicians even noticed that there was interesting mathematics here. Instead, the patterns were developed by artisans, probably by experimentation. It was only in the 19th and 20th centuries that mathematicians discovered the underlying mathematical concepts.

The Arabic tradition was intensely creative, picking up the best from Greek and Indian mathematics and developing it further. Unfortunately, only a small part of it was transmitted to Europe. As a result, many of these results had to be rediscovered, sometimes many centuries later. It was in the 19th century that European scholars began to investigate the Arabic mathematical texts. Since then, historians have learned a lot about this period. There are still many manuscripts to read and study, and new discoveries are being made all the time. Our picture of the Arabic achievement is still incomplete.

To read more about Arabic mathematics, one could start with [74, Section 1.6], [99, Chapter 7], and [157, pp. 137–165]. Some of the articles in [170, Part V] offer specific examples of the work of Arabic mathematicians. Two useful book-length introductions are [13], which is more mathematical, and [150], which is more interested in astronomy. See also [144, Volume 2].

Medieval Europe

Around the 9th century, political and social life in Western Europe began to be stable enough for people to pay attention to education. In many places "cathedral schools" sprang up, dedicated to the training of future priests and clerics for the local diocese.

They concentrated on the ancient tradition of the introductory *trivium* (grammar, logic, and rhetoric). The more advanced students might go on to the *quadrivium* (arithmetic, geometry, music, and astronomy). It is likely that very few students actually studied the mathematical topics in the quadrivium, but their presence in the curriculum helped stimulate interest in mathematics.

Once people became interested in mathematics, where could they go to learn more? The obvious thing was to go to places under Islamic control, of which the most accessible was Spain. Gerbert d'Aurillac, later to be Pope Sylvester II, is an example. Gerbert visited Spain to learn mathematics, then reorganized the cathedral school at Rheims, France. He reintroduced the study of arithmetic and geometry, taught students to use the counting board, and even used the Hindu-Arabic numerals (but not, it seems, the full place value system).

In the following centuries, many European scholars spent time in Spain translating Arabic treatises on all sorts of subjects. While few European scholars knew Arabic, many Jewish scholars living in Spain did, so the translations were often done by having a Jewish scholar translate from Arabic to some common language, then translating from that language into Latin. Many of the mathematical and philosophical texts in Arabic were translated. In addition, a wide range of ancient Greek texts, from Aristotle (the most influential) to Euclid, were translated from the Arabic to make their impact in the West.[16]

Spain is very far from Baghdad, and it was not really a center of mathematical activity. Only the oldest and easiest mathematical texts

---

[16]The table in [99, p. 291] gives a good overview of this translation enterprise.

were likely to be found in Spanish libraries. This is probably the reason why al-Khwārizmī's works were so prominent. His algebra book was translated by Robert of Chester in 1145. (It was this translation that Latinized "al-jabr" into "algebra.") His book on arithmetic with Hindu-Arabic numerals seems to have been translated or adapted several times. Many of these versions began with the words "dixit Algorismi" ("so says al-Khwārizmī"), and it is because of this expression that the word *algorism* came to mean the process of computing with Hindu-Arabic numerals. The modern word *algorithm*, meaning a "recipe" for doing something, is a modern variant of *algorism*, which was a recipe for doing arithmetic.

The system of cathedral schools eventually led to the establishment, in the 11th and 12th centuries, of the first universities in Bologna, Oxford, Paris, and other European cities. For the most part, the scholars at the universities were not interested in mathematics. Aristotle's work did have a great impact, however. His work on the theory of motion led a few scholars at Oxford and Paris to think about *kinematics*, the study of moving objects. It is to these scholars that we owe the notions of instantaneous velocity and of uniformly accelerated motion. Perhaps the greatest of them was Nicole Oresme, at the University of Paris. Oresme worked on the theory of ratios and on several aspects of kinematics, but his most famous contribution is a graphical method for representing changing quantities that anticipates the modern idea of graphing a function. In addition, questions of motion led him to consider infinite sums of smaller and smaller terms. He obtained several significant results about such sums, sometimes by using ingenious graphical tricks.

Trade was another source of contact between Europeans and the Islamic Empire. Leonardo of Pisa[17] was the son of a trader. Traveling with his father, he learned quite a bit of Arabic mathematics. In his books, Leonardo explained and extended what he had learned. His first book was the *Liber Abbaci* ("Book of Calculation"), published around  1202 and revised in 1228. It started by explaining Hindu-Arabic numeration and went on to consider a wide array of problems. Some of the problems were practical: currency conversion and computing profits, for example. Others were more like the word problems in today's algebra texts. The book contains a geometric explanation of the rules

---

[17]Leonardo is often referred to as "Fibonacci," a reference to his father's name. There is no evidence, however, that Leonardo ever used this name for himself.

for solving quadratic equations and a few other theoretical passages, but the focus is on the problems and the methods for solving them.

Leonardo's other works are also important. His *Practica Geometriae* is a manual of "practical geometry" which seems to have been heavily influenced by the work of Abraham bar Hiyya, who lived in Spain in the 12th century. (Abraham bar Hiyya wrote in Hebrew, but his work was one of many translated into Latin during that century.) Leonardo also wrote the *Liber Quadratorum*, the "book of squares." Here he reveals himself to be a creative and talented mathematician. The book discusses how to solve various kinds of equations involving squares, under the restriction that the solutions be whole numbers. The problems typically involve more than one variable. The *Liber Quadratorum* anticipates and points toward the work of Fermat and Euler, 400 or more years down the road.

Leonardo's work may have been one of the sources of what was to become a lively tradition in Italy. As the Italian merchants developed their businesses, they had more and more need of calculation. The Italian "abbacists" tried to meet this need by writing books on arithmetic and algebra. They often wrote in everyday Italian rather than in Latin, which was the language of the scientists. The culmination of this tradition was the work of Luca Pacioli, whose *Summa de Aritmetica, Geometria, Proportione e Proportionalita* was a huge compendium of practical mathematics, from everyday arithmetic to double-entry bookkeeping. Printing had just been developed in Germany and was flourishing in Italy. Pacioli's *Summa* was one of the first mathematics

 books to be printed. This gave it wide circulation, and it became the basis of much later work on algebra. (See Sketches 2 and 8 for details about Pacioli's notation.)

Not much has been written specifically about the history of mathematics in Medieval Europe, though many translations and editions of the main texts are available. There is a good overview of this period in [99, Chapter 8]; part 2 of [74] is a useful complement. Biographies are a good way to get a sense of the times: Gerbert d'Aurillac is the topic of [18], while Leonardo of Pisa is at the center of [46] and [45].

The 15th and 16th Centuries

Around the end of the 14th century, many different cultures around the world were producing interesting mathematics. In Central America, the Maya had developed a base-twenty system for numeration (see

Sketch 1) and an elaborate calendar. In China, mathematicians had developed sophisticated methods for solving many kinds of problems. Both Indian and Arabic mathematics had also continued to grow.

These cultures were somewhat insulated from each other and from European culture. There were some contacts, especially in commerce, but little mathematical knowledge seems to have been exchanged at that time. All of that was about to change dramatically. Beginning in the 15th century and intensifying from then on, Europeans began to develop the art of navigation and to travel to distant continents, taking European culture with them. By the late 1500s, Jesuit schools had been established in many places, from South America to China. The Jesuit cultural network eventually extended throughout the world. As a result, European mathematics was taught and studied everywhere and eventually became the dominant form of mathematics worldwide.

As European sailors began to travel to other continents, solving the technical problems of navigation became more and more important. Long-range navigation depends on astronomy and on a good understanding of the geometry of the sphere; this helped propel trigonometry to the center of atten-  tion. Astrology was also a very important part of the culture of this period, and making star charts also depends on having a good grip on (spherical) trigonometry. Because of this, trigonometry was one of the major themes in the mathematics of the 15th and 16th centuries.

In parallel with the intense study of navigation, astronomy, and trigonometry, there was also growing interest in arithmetic and algebra. With the rise of the merchant class, more people found that they needed to be able to compute. Since algebra was thought of as a kind of generalized arithmetic, it was natural for scholars to move from arithmetic to algebra as they went deeper into their studies. Algebra remained a central interest of mathematicians well into the 17th century; we discuss it in detail in the next section.

Of course, algebra and trigonometry are related, and they influenced each other. Trigonometry is a kind of algebraized geometry, and both algebra and trigonometry are methods for solving problems. Often the same scholars wrote on both subjects. A leading example was Johannes Müller, also known as Regiomontanus (a Latinized version of "from Königsberg," referring to his birthplace). Besides translating many classical Greek works and studying the stars, he wrote *De Triangulis Omnimodis* ("On All Sorts of Triangles"), one of the first treatises devoted solely to trigonometry.

A great many new ideas were introduced into trigonometry at this time. The list of trigonometric functions (sine, cosine, tangent, cotangent, secant, cosecant) became standardized. New formulas and new applications were discovered. Given the interest in navigation and astronomy, most of the focus was on spherical triangles, whether on the celestial sphere or on the earth. Throughout all this, sines and cosines continued to be thought of as lengths of particular line segments. No one thought of them as ratios or as lengths on a unit circle. All sine tables were based on a circle of fixed radius, and in applications one had to use proportionality to adapt the information to the radius at hand. (See Sketches 18 and 26 for more about early trigonometry.)

Somewhat related to all this was the discovery of perspective by Italian artists. Figuring out how to draw a picture that gave the impression of three-dimensionality was quite difficult. The rules for how to do it have real mathematical content. Though the artists of the Renaissance did not subject these rules to a complete mathematical analysis, they understood that what they were doing was a form of geometry. Some of them, such as Albrecht Dürer, were quite sophisticated in their understanding of the geometry involved. In fact, Dürer wrote the first printed work dealing with plane curves that went beyond the conics, and his investigation of perspective and proportion is reflected both in his paintings and in the artistic work of his contemporaries. (See Sketch 20.)

There are good surveys of the mathematics of this period in [99, Chapters 9 and 10] and [74, Part 2]. For the non-European aspects, look at [157]. The best account of the early history of trigonometry is [175]. For perspective, look at [61] and [6].

Algebra Comes of Age

As we reach early modern times, mathematics begins to get broader and more diverse. While certain concerns were dominant at certain times, there were always other topics being studied. From this point on, our overview selects and follows a few important threads, rather than attempting to cover everything. In the 16th and early 17th centuries, it is algebra that takes center stage.

It's important to understand what algebra was like at this time. Leonardo's *Liber Abbaci*, like the Arabic algebra which inspired it, was entirely rhetorical. Equations and operations were expressed in words, written out in full. Between the time of Leonardo and the 16th century, scholars developed abbreviations for many of the words, such as writing *p* for "plus." But they didn't really change the nature of the enterprise and, in particular, did not introduce any sort of general or standardized notation. One started from the rules of arithmetic and attempted to solve equations. The equations treated were mostly of degree 1 or 2 or were easily reducible to such equations.

The Italian algebraists used the word *cosa*, meaning "thing," for the unknown quantity in their equations. When scholars from other countries got involved, they used *coss*. As a result, they were sometimes called the "cossists," and algebra was sometimes known as "the cossic art."[18] The English scholar Robert Recorde is a good example of this tradition. In his *The Grounde of Artes* (1544), he explains basic arithmetic. This book, written in English in dialogue form, was so popular that it went through 29 editions. His next book, *The Whetstone of Witte* (1557), is a kind of sequel which he describes as dealing with "the cossike arte," that is, algebra. (The title is a pun: *cos* happens to mean "whetstone" in Latin!) This was typical of the early algebraic tradition: it was viewed as a popular, practical kind of mathematics, a way to sharpen your wit after having laid down the groundwork of arithmetic.

The whetstone of witte, whiche is the seconde parte of Arithmetike: contaynyng the extraction of Rootes: The Cossike practise, with the rule of Equation: and the woorkes of Surde Nombers.

[from Recorde's title page]

Recorde in England is just one example. From Pedro Nunes in Portugal to Michael Stifel in Germany, the cossists were all over Europe. They were particularly good at inventing notations. They had different symbols for the unknown, the square of the unknown, and so on. Some used special notations for the algebraic operations and for the extraction of roots. Others simply abbreviated the crucial words. They did not use symbols for quantities other than the unknown; these were always numerical. Hence, they could write the equivalent of $x^2 + 10x = 39$, but they could not write something like $ax^2 + bx = c$. Thus, while they certainly "knew" the quadratic formula, they could not write it down as a formula the way we do. Instead, they would give a verbal recipe and many examples. (See Sketches 2 and 8 for more about the evolution of algebraic symbolism.)

Several algebraists of this period attempted to find a method to

---

[18]English spelling was not standardized at that time, so *cossic* also appears as "cossick" or "cossike" in various places.

solve cubic equations. The crucial breakthrough was made in Italy, first by Scipione del Ferro and then by Tartaglia.[19] Both men discovered how to solve certain kinds of cubic equations. They kept their solutions secret, because at this time scholars were mostly supported by rich patrons and had to earn their jobs by defeating other scholars in public competitions. Knowing how to solve cubic equations allowed them to challenge the others with problems that they knew the others could not solve, so people were inclined to keep quiet about their discoveries.

In the case of the cubic, this pattern was broken by Girolamo Cardano. Promising never to reveal it, Cardano convinced Tartaglia to share the secret of the cubic with him. Once he knew Tartaglia's method for solving some cubic equations, Cardano was able to generalize it to a way of solving any cubic equation. Feeling that he had actually made a contribution of his own, Cardano decided that he was no longer bound by his promise of secrecy. He wrote a book called *Ars Magna* ("The Great Art"). This

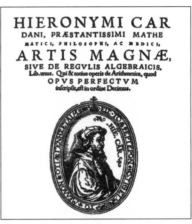

[from Cardano's title page]

scholarly treatise (written in Latin) gave a complete account, with elaborate geometric proofs, of how to solve cubic equations. It also included a solution for the general equation of degree 4 that had been found by Cardano's student Lodovico Ferrari. Despite the fact that Cardano acknowledged his contribution, Tartaglia was furious at Cardano's breach of promise. He protested publicly, but in the end there was little he could do. In fact, the formula for the solution of a cubic equation is still often called "Cardano's formula." (See Sketch 11 for more about this story of human nature at work.)

Cardano's notation was still in the old tradition. He would have written the equation $x^3 = 15x + 4$ something like this:

cubus.aeq.15.cos.p.4,

which was read as "a cube is equal to 15 things plus 4." He would then give a procedure for solving the equation that was basically the same as the cubic formula we use today. Since the coefficients of his

---

[19]Though this mathematician's real name was Niccolò Fontana, everyone knows him by his nickname, Tartaglia, which means "stammerer."

equations were always numbers, he described the method in words and gave examples of its application.

There was one very serious problem, though. Applying Cardano's method to the equation above yields (in modern notation)

$$x = \sqrt[3]{2 + \sqrt{-121}} + \sqrt[3]{2 - \sqrt{-121}}.$$

In a quadratic equation, the appearance of the square root of a negative number is a signal that there is no solution. But here it's easy to check that $x = 4$ is a solution. Cardano's methods solved most cubics, but got him in trouble in cases like this. His solution to this issue was to sweep it under the rug. Equations like this one are barely mentioned in this book, and the few mentions are cryptic and short. (See Sketch 17.) His method worked for every one of the equations he discussed in detail.

The missing step was provided by Rafael Bombelli. In his *Algebra*, Bombelli extended Cardano's ideas in several different directions. In particular, he developed a method for dealing with cubic equations which led to expressions such as the one above, involving square roots of negative numbers. We might describe Bombelli's method as the beginning of a theory of the complex numbers, though it is unclear whether he really thought of them as "numbers." He described them as "linked cubic radicals of a new type" and explained how to operate with them in such a way as to obtain the solution. (See Sketch 17 for more about the beginnings of complex numbers.)

Another important idea developed by Bombelli was to link algebra and geometry in a more direct way. Cardano could only conceive of x^3 as meaning the volume of a cube of unknown side. Bombelli, on the other hand, was prepared to consider diagrams containing segments whose length was, say, x^2 and to use these diagrams to explain his equations. By emphasizing the connection to geometry, Bombelli was anticipating the direction algebra would soon take.

Bombelli's *Algebra* had one other important feature. While he was working on the book, Bombelli learned that Regiomontanus had translated a portion of Diophantus's *Arithmetica*. He located it and read it, and it is clear that this old Greek book had a huge impact on his thinking. In fact, a large portion of the latter part of Bombelli's work contains problems taken from Diophantus. These problems were usually equations in several variables which had many solutions, but for which one was supposed to find solutions in whole numbers or in fractions. They require a completely different set of algebraic techniques. Bombelli didn't go much beyond what Diophantus had already done, but here, too, he paved the way for future developments.

Algebra began to look more like it does today towards the end of the 16th century, in the hands of François Viète. Among many other things, Viète worked for the French court as a cryptographer, a code-breaker who deciphered intercepted secret messages. This may be what led him to one of his most important innovations: the notion that one  could use letters to stand for numbers in equations. Viète used the vowels A, E, I, O, U to indicate unknown quantities and consonants to indicate known numbers. He was thus the first to be able to write an equation like $ax^2 + bx = c$, though his version would have looked something like

$$B \text{ in } A \text{ squared } + C \text{ plane in } A \text{ eq. } D \text{ solid.}$$

(Remember that the vowel A is an unknown and the consonants B, C, and D stand for numerical parameters).

The form of this equation highlights one of Viète's worries: he always wanted quantities to be equidimensional. Since multiplying B by A^2 would make a "solid" (i.e., three-dimensional) quantity, he specified that C must be plane, so that CA is also solid, and that D must be solid. Bombelli, as mentioned above, had no such worries. Descartes would later convince everyone that it was best not to worry about dimensionality.

Perhaps the most important thing Viète did, however, was to promote algebra as an important part of mathematics. In algebra, one often starts by assuming that, say, $ax^2 + bx$ is, in fact, equal to c, and then one deduces what x must be equal to. This reminded Viète of what the 4th century Greek mathematician Pappus called *analysis*, in which one assumes a problem has been solved and proceeds to deduce  things from this assumption. Hence, Viète argued that the Greek analysis was actually algebra, which the Greeks were supposed to have known and kept secret.[20] At this time, algebra was often considered less serious and less important than geometry. By giving it a Greek pedigree, Viète made algebra more acceptable to more mathematicians. (See Sketch 8 for more about Viète's contributions to algebra.)

René Descartes completed the process of bringing algebra to a mature state. In his famous book *La Géométrie* (published as an appendix to *Discourse on Method*), Descartes proposed essentially the notation

---

[20]He was wrong about this, but inventing a Greek origin for algebra was useful at a time when classical learning was greatly valued.

we use today. He suggested the use of lowercase letters from the end of the alphabet (such as x, y, and z) for unknown quantities and lowercase letters from the beginning of the alphabet (such as a, b, c) for known quantities. He also came up with the idea of using exponents on a variable to indicate powers of that variable. Finally, he pointed out that once one fixes a unit length, any number can be interpreted geometrically as the length of a line segment. So x^2 could just mean a line segment whose length is equal to the number x^2. With this, Descartes did away with Viète's worries about dimensionality. (See Sketch 8 for more about how Descartes influenced algebraic notation.)

Three innovations from this period were to be extremely important. First, the fact that no one could figure out how to solve the general quintic (fifth-degree) equation led algebraists to start asking deeper questions. Slowly, a theory about polynomials and their roots evolved. Second, Descartes and Pierre de Fermat linked algebra and geometry, inventing what we now call "coordinate geometry." They showed that, just as we can interpret algebraic equations geometrically, we can also interpret geometric relations algebraically. (See Sketch 16 for more about coordinate geometry.) Both Fermat and Descartes demonstrated the power of algebra to solve geometric questions by applying it to famous difficult geometric problems that had been discussed by Pappus. Third, Fermat introduced a whole new category of algebraic problems. These were related to the work of the Greek mathematician Diophantus, but went far beyond his work. Specifically, Fermat began asking "questions about numbers," by which he meant *whole* numbers. For example, can a square be equal to 1 plus a cube? This amounts to the equation $y^2 = x^3 + 1$, which Fermat wanted to solve under the restriction that x and y should be whole numbers. There is an easy solution, $x = 2$ and $y = 3$. But are there any others? Fermat developed methods for answering such questions. He could also use his methods for proving what he called "negative propositions." These were statements that said that certain equations *cannot* be solved. For example, he proved that the equation $x^4 + y^4 = z^2$ has no solution in (non-zero) whole numbers.

Unfortunately, for a long time Fermat was alone in finding these questions interesting. Mathematicians of the time felt that good mathematics was about what *can* be done, so their reaction to Fermat's negative propositions was puzzlement. Why glory in not being able to solve certain problems? Furthermore, Fermat was not a professional mathematician, but a lawyer serving on the court in Toulouse, France. As a result, he never published his proofs. Instead, he wrote letters to friends, explaining his ideas and discoveries, but without the technical details. It was left to other mathematicians, a century later, to rediscover the subject and find the proofs all over again. We'll get back to that when we discuss the work of Euler and others in the 18th and 19th centuries.

The early history of algebra is discussed in Chapters 9 and 11 of [99]. For a more detailed treatment, one might go to [158] or [100].

Calculus and Applied Mathematics

While the mathematicians of the late 16th and early 17th century were developing algebra, another group of people were beginning to use mathematics to try to understand the universe. At that time, they were usually called "natural philosophers." Perhaps the most famous of them was Galileo Galilei, who lived mostly in Florence, Italy. Galileo's studies included astronomy and the physics of moving bodies. He blended observation and experimentation with mathematical analysis. In fact, he insisted that one had to use mathematics in order to have a chance of understanding the world.

He was not alone. Johannes Kepler, in Germany, was using the old Greek geometry of conic sections to describe the solar system. The Greeks had studied the ellipse for its mathematical interest. Kepler discovered that the planets moved around the sun in elliptical orbits and formulated mathematical laws to describe how fast each planet moved. (See  Sketch 28.) In France, Father Marin Mersenne,[21] was trying to get many scholars from different areas to come together, talk, and cooperate in order to understand the world. In England, Thomas Harriot was developing algebra and applying mathematics to optics, navigation, and other questions. And René Descartes, who had revolutionized both algebra and geometry by bringing them together, also began trying to understand comets, light, and other phenomena.

All of this work brought some specific mathematical issues to everyone's attention. Studying motion inevitably led to difficult questions related to the infinite divisibility of space and time. When an object is moving in such a way that its velocity is continually changing, how does one even understand what its velocity is? And how does one figure out what distance it covers in a given amount of time?

---

[21]Marin Mersenne was a Franciscan friar, an important music theorist, and a competent amateur mathematician.

These questions were not new. They had been studied and discussed by medieval scholars such as Nicole Oresme and by many others since then. But their new relevance made them more urgent. So did the related questions of finding tangents to curves and areas of curved figures. As a result, many mathematicians started to work on them. One such person was Bonaventura Cavalieri, a Jesuat[22] former student of Galileo and professor at the University of Bologna. To study the areas and volumes of curved figures, Cavalieri worked with the "principle of indivisibles." This is the idea that a planar region can be considered as an infinite set of parallel line segments and a solid figure can be considered as an infinite set of parallel planar regions. For example, a cylinder would be thought of as a stack of circles, all of the same size, and a sphere would be a stack of circles of varying sizes. Using this idea, Cavalieri could compute many areas and volumes that would previously have been very hard
to analyze. Similar work was done by Fermat and Descartes in France and by many other European mathematicians. By 1660 or so, it was clear that problems of this kind could be solved as needed. What was not available, however, was a general method.

In the late 1660s, Isaac Newton and Gottfried Wilhelm Leibniz independently discovered such a method. In fact, they discovered two slightly different methods. Newton's approach emphasized what he called "flowing quantities" and their rates of flow, which he called their "fluxions." Leibniz's approach used the idea of "infinitesimal," or infinitely small, quantities. If a quantity was represented by a variable x, Leibniz defined its "differential" dx to be the amount by which it changed in an infinitesimal amount of time. Both fluxions and differentials are basically what we now call *derivatives*. (See Sketch 30.)

The most important part of the discovery may well have been the realization that one could develop a kind of recipe for computing these things. It was a method of calculation, "a calculus." Leibniz was the one who saw this most clearly. In his very first paper about the subject, he emphasized the importance of being able to solve problems without really having to *think* about what was going on. One simply applied the rules of the calculus.[23]

---

[22]Not to be confused with "Jesuit." The Jesuati were a congregation within the Catholic Church from 1366 to 1668.

[23]Those of us who teach calculus today may be forgiven for feeling that Leibniz did the job a bit too well!

With such a powerful tool in hand, many people set out to apply it to the world. This became the main theme of 18th century mathematics. Newton himself had prepared the way with his *Principia*, a mathematical analysis of the laws of motion and the workings of the solar system. Jakob Bernoulli, his brother Johann, and others learned to use the calculus. Johann played an especially important role. Since his brother was professor of mathematics in their home town (Basel, Switzerland), he had to go elsewhere. To earn money, he arranged to teach the new calculus to a French nobleman, the Marquis de l'Hospital. They agreed that Bernoulli would write letters explaining the calculus, and the content of these letters would then belong to l'Hospital. The result was the first calculus textbook, published by l'Hospital in 1696.

Soon there were many such books, and many mathematicians were using the calculus to study the natural world. Among them was Daniel Bernoulli, son of Johann Bernoulli, who studied mechanics and hydrodynamics. In Italy, Maria Gaetana Agnesi became one of the first women to have an impact on modern mathematics when she wrote her own textbook, a unified treatment of algebra, coordinate geometry, and calculus. Agnesi's book was rather old-fashioned, but it proved useful to many people. Agnesi herself became quite well known; she was even named professor at a university, though she did not accept the position.

In France, the spread of the new mathematical ideas of calculus was tied to the spread of the new Newtonian ideas about motion and gravitation. Many scholars were involved in this process. One of the most important was a woman, Emilie de Châtelet. Among other things, she produced a richly annotated translation into French of Newton's *Principia*. Since Europeans tended to prefer the language and notation used by Leibniz, Châtelet also had to translate the mathematics. In the process she clarified several significant issues and helped convince people that the new physics was correct.

The greatest mathematician of the time, however, was Leonhard Euler. He was born in Switzerland and had private lessons from Johann Bernoulli (who by this time had replaced his older brother on the faculty at Basel). Euler spent most of his life, however, in St. Petersburg (in Russia) and Berlin (then the capital of Prussia). At this time, the universities had little interest in scientific research. Instead, several of the European kings established royal scientific academies where scholars could work and communicate. There was even some competition as to who could catch the best scholars. For almost all of his life,

Euler was associated with the St. Petersburg Academy. At one point, he was lured away and spent a few years at the Berlin Academy, but he ended up returning to Russia.

Reading Euler's work gives one the impression of a mind overflowing with ideas. He wrote on every aspect of mathematics and physics and also had much to say about astronomy, engineering, and philosophy. In mathematics, he developed the calculus into a powerful instrument and applied it to all sorts of complicated problems, both in pure mathematics and in physics. He wrote textbooks explaining all this to others. In his "precalculus" textbook, he emphasized the idea of a function. Many of the conventions and notations we still use were introduced in his books.

Euler didn't stop there. He rediscovered Fermat's number theory and put it in order. He found correct proofs of Fermat's statements and established number theory as an important part of mathematics. He investigated algebra and polynomials and came close to proving the Fundamental Theorem of Algebra. He studied the geometry of the triangle, discovered a basic theorem about solid polyhedra, and began to study the geometry of curves and surfaces. He applied mathematics to the design of ships and turbines and to other engineering problems. He even considered lotteries and a puzzle about walking over a set of seven bridges without crossing any of them twice. When the mathematics needed to study something was available, he used it. When it wasn't, he developed it. Euler's works now fill more than 80 printed volumes. Euler didn't wait until he had just the right approach to analyzing something. He just plowed in and got to work. If later he saw a simpler or better way, he'd write another paper. Reading his papers allows one to see why he was interested in each problem and to follow how his ideas developed.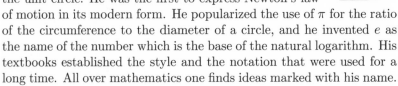

Euler's influence was enormous. He was the first to suggest that it was best to consider the sine and cosine as functions of the angle, and to define them in terms of the unit circle. He was the first to express Newton's law of motion in its modern form. He popularized the use of π for the ratio of the circumference to the diameter of a circle, and he invented e as the name of the number which is the base of the natural logarithm. His textbooks established the style and the notation that were used for a long time. All over mathematics one finds ideas marked with his name.

Perhaps one can see the culmination of 18th century mathematics in the work of two mathematicians, Pierre Simon Laplace and Joseph-Louis Lagrange. Laplace was an applied mathematician through and through. He wrote famous books on celestial mechanics and on probability. They are both huge books, dense with mathematics. They also both come accompanied by more popular versions: *The System of the World* to explain how the solar system works, and a *Philosophical Essay on Probabilities* that explained the basic ideas and argued for their broad applicability. (See Sketch 21 for the early history of probability.) Lagrange, for his part, worked in all areas of mathematics, from algebra to the calculus. One of his most important books is a treatise on mechanics, in which he lays out a mathematical theory of how and why things move. This remarkable book is quite important in the history of physics, but it is also famous for having no pictures in it at all! Lagrange felt that his formulas had captured physical reality so well that diagrams were no longer needed.

The 18th century was an optimistic time. Laplace did not hesitate to apply his calculus of probabilities to the activities of judges and juries. Modern readers often find his assumptions very strange.[24] Similarly, Lagrange did physics without diagrams. Mathematicians felt they had in their hands the key to reality. And, for the most part, the key did seem to unlock many doors.

Not all was well, however. The most famous sign of trouble was a tract by George Berkeley, Irish philosopher and Anglican Bishop of Cloyne. Berkeley was offended by the public atheism of a mathematician who argued that theology was speculative and uncertain. He responded by showing that the foundations of the calculus, which the mathematicians prized so much, were just as shaky. His scathing essay, *The Analyst*, showed that both Newton's fluxions and Leibniz's infinitesimals were ill-defined and in some ways contradictory. Fluxions were computed by taking the quotient of two increments. Newton seemed sometimes to say these increments were both zero and sometimes to say that they were not. Berkeley put his finger directly on the contradiction:

> And what are these fluxions? The velocities of evanescent increments. And what are these same evanescent increments? They are neither finite quantities, nor quantities infinitely small, nor yet nothing. May we not call them the ghosts of departed quantities?

The central point is made early in the essay: "... he who can digest a second or third fluxion... need not, methinks, be squeamish about any point in Divinity."

---

[24] Is it really reasonable, for example, to think of human decision-making as analogous to drawing a colored ball from an urn?

Berkeley also argued that Newton, for example, would cut corners in his writing, skipping over difficulties or choosing just the right way to say something so that certain issues were obscured. Berkeley's critique showed that the very foundations of the calculus were insecure.

For many mathematicians, the answer was "but it works!" Jean le Rond d'Alembert, a French mathematician and philosopher who was one of the editors of the famous *Encyclopédie*, is supposed to have said to a student: "Persist, and faith will come." Nevertheless, several mathematicians, including D'Alembert himself, began to look for better foundations. The first results of this search came early in the 19th century, though the task was not completed until much later.

The early history of the calculus has been the subject of much historical investigation. Chapters 12 and 13 of [99] are a good place to start, and the first half of [75] is a good source of further information. Both contain pointers to further reading.

Rigor and Professionalism

The 19th century saw a huge explosion in mathematical activity and significant changes in where and how mathematicians did their work. Progress was made in so many different directions that it is hard to find one main theme which characterizes all this work. We've chosen to emphasize three important aspects of 19th century mathematics. First, there was a deep concern with rigor, especially with regard to calculus. Second, physical problems led to more, and more sophisticated, mathematics. Third, mathematicians became professionals in a new and different way.

The opening years of the 19th century were marked by the aftermath of the French Revolution of 1789. The revolution affected all of Europe. Among the changes the revolutionaries introduced in France was a new emphasis on education. They established schools, such as the École Polytechnique in Paris, whose main mission was to provide a technical education to the middle class. The goal was to create a class of well-trained civil servants to run the new French Republic. One result was that mathematicians were now expected to teach. What's more, their students were expected to learn from their teaching. This put a new value on clarity, precision, and rigor. After all, if the teacher doesn't understand the basis of what is being studied, how can the student understand it? At about the same time, the French Academy of Sciences devised the *metric system* as a

standardized system of measurement for use in science and commerce. After its official adoption by the French government in 1795, this system gradually spread to other countries. By the end of the 19th century, it had become a recognized international standard. (See Sketch 6.)

As the 19th century opened, the dominant mathematical figure was Carl Friedrich Gauss. He was a child prodigy, able to do arithmetic by the time he was three. At age 17, he was already making significant new discoveries, which he recorded in his mathematical diary. His first important book, published in 1801, was called *Disquisitiones Arithmeticae* (Arithmetical Investigations). It dealt with whole numbers and their properties, and it was marked by what came to be known as the Gaussian style: spare, precise, with almost no motivation or explanation beyond the technical proofs.

Gauss's work spanned all of pure and applied mathematics. In fact, he is just as famous for his work on physics and astronomy as he is for his strictly mathematical work. He also had a way of combining the two. For example, after doing some surveying, he applied ideas he had developed for that task to his study of the geometry of surfaces. This interaction between applications and theory was quite typical of the work of the great 19th century mathematicians.

The other great figure of the early 19th century was Augustin Louis Cauchy. As a very young man, Cauchy was made a professor at the École Polytechnique in Paris. His main job was teaching calculus, which he preferred to call *analysis*. He clearly was not satisfied with the way the foundations of the subject then stood. The result was one of the most famous textbooks in the history of mathematics. The title, *The École Polytechnique Course in Analysis*, didn't really indicate how revolutionary the contents were. The first volume[25] was called *Algebraic Analysis* and tried to make precise such ideas as continuity and convergence. In his courses, Cauchy's goal was to "do calculus right." For the first time, there were *definitions* of the derivative and the integral. For the first time, the Fundamental Theorem of Calculus was highlighted as indeed fundamental. And, much as we do today, Cauchy emphasized the algebraic side of calculus, preferring computations to diagrams and formulas to geometric intuitions.

Cauchy's work ranged far and wide in mathematics and mathematical physics. He wrote about everything. In fact, he was so prolific that at one point the editors of a major French journal imposed a quota on him. In response, he convinced a publisher (who happened to be family) to put out a journal containing nothing but papers by Cauchy! His

---

[25]There never was a second.

emphasis on rigor had a great impact on other mathematicians. There is a story that Laplace once listened to him explain the importance of *convergence* of infinite series. Alarmed, Laplace ran home to check the series in his massive book on celestial mechanics. The influence of the new ideas was deep.

Many mathematicians continued Cauchy's work of making the calculus precise and rigorous. In the process, they found flaws in his work, of course, and tried to improve on it. The most important of these mathematicians was Karl Weierstrass, whose name, even today, is synonymous with rigor and precision. Weierstrass was also a teacher. In fact, he began his career as a high school teacher. Besides mathematics, he also had to teach other subjects, from physics and botany to calligraphy and gymnastics. It was only after publishing several scientific papers that he was recognized as a creative mathematician. He eventually became a professor at the University of Berlin.

Weierstrass's analysis lectures in Berlin were famous. He was not an animated teacher. Health problems forced him to lecture sitting down while a student wrote formulas on the blackboard. But the content of the lectures was inspired, and many of his students became great mathematicians. One achievement of these lectures was a complete transformation of the basis of calculus. His clear, precise definitions removed any trace of mystery or geometric intuition from calculus, putting it all on a logical foundation that depended only on algebra and arithmetic. The new approach wasn't easy, though, as students who have had to learn his "epsilon-delta" approach to limits will still testify.

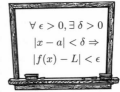

Having settled the issue of finding rigorous foundations for the calculus, mathematicians considered other parts of mathematics, too. Richard Dedekind and Giuseppe Peano investigated the foundations of arithmetic, and Georg Cantor invented the notion of a set, which allowed him to make fundamental discoveries about infinity. The theory of sets eventually came to be seen as a possible foundation for all of mathematics. (See Sketch 25.)

Algebra and geometry also were changed in fundamental ways in the 19th century. Algebraic formulas for the solution of equations of degrees 3 and 4 had been found in the 16th century, but no one had been able to solve the equation of degree 5. Gradually, mathematicians had begun to turn from the search for a solution to the search for understanding of why the solution was not forthcoming. In the 18th century, Joseph-Louis Lagrange had noticed that one could understand all the existing solutions by analyzing the effect of permuting the roots of an equation on various polynomial expressions. This showed mathematicians how to think more rigorously about what a "formula" for solving an equation actually is. Together with the discovery of the connection between formulas and permutations, this eventually led to the fundamental algebraic discoveries of the 19th century.

The first breakthrough came from Niels Henrik Abel, a Norwegian mathematician whose brilliant career was cut short by an early death. Abel managed to prove in 1822 that in fact *there is no general formula* for the solution of the equation of degree 5. This was a remarkable development, but the real revolution in algebra came from the work of Évariste Galois. Galois was brilliant and temperamental, passionate about both mathematics and politics. As a young man, he spent a large part of his short life being thrown out of schools and into jails. Despite these frequent embroilments, he devoted much of his time to mathematics, but his writings went unnoticed. In 1832, shortly before his twenty-first birthday, Galois became involved in a duel that cost him his life.

The night before the "affair of honor," he dashed off a hurried letter to a friend, summarizing the discoveries that he had made in several mathematical papers. He concluded the letter by asking his friend to send it to the best mathematicians of the day and expressing the hope that later there would be "some people who will find it to their advantage to decipher all this mess." He was right. "This mess" was the analysis of the solvability of algebraic equations by means of group theory, a subject which has become the cornerstone of modern algebra and geometry.

Galois also introduced a fundamental change of perspective. He argued that it was necessary to move from considering particular transformations of equations to considering all possible transformations at once. This move to a more abstract perspective was picked up by other mathematicians as the century progressed and eventually led to what is now known as "abstract algebra."

In geometry too, it was a time of revolution. For centuries, people had thought about Euclid's parallel postulate and its role in plane geometry. The work of Gauss, János Bolyai, Nicolai Lobachevsky, and Bernhard Riemann finally settled the issue, leading to the discovery of the non-Euclidean geometries. (See Sketch 19.) Once again, the move was toward abstraction and rigor: rather than attempting to figure out what was "the geometry of the real world," these mathematicians showed that there were several alternative ways to do geometry, each consistent within itself, each interesting, each correct. It must have seemed, at the time, to be a useless but beautiful dream, a move away from applicability. But it didn't turn out that way. When Albert Einstein, in the 1910s, was looking for a way to express his insights into gravitation, he found the right language in Riemann's approach to geometry. It turns out that we may live in a non-Euclidean universe!

The connection between mathematics and theoretical physics remained strong. Applied mathematics was a source of interesting and difficult problems. To solve them, important new mathematics had to be created. In order to study how heat spreads through objects, Joseph Fourier invented what are now known as "Fourier series." These proved to be very useful in applied mathematics, to study light, sound, and other periodic wave phenomena. They also turned out to be interesting mathematical objects, and unraveling their properties required the full power of the rigorous approach to calculus that had just been developed. Many other areas of physics were studied, too. Electricity and magnetism posed interesting problems that led to important mathematics. Understanding machines, fluid flow, the motion of the planets, the stability of structures, the tides, the behavior of elastic materials — all this engaged the attention of many mathematicians.

One example is Sophie Germain, a contemporary of Gauss, Cauchy, and Fourier. Despite growing up during the so-called "Enlightenment," Germain faced real resistance to her desire to become a mathematician. Women could not study at the *Polytechnique*, so Germain had to borrow notes from male students in order to be able to learn advanced mathematics. Overcoming her society's prejudice against female intellectuals, she became one of France's finest mathematical scholars. Besides doing significant work in number theory (see Sketch 13 for more information), Germain obtained important results in the mathematical theory of elastic materials. Many mathematicians built on her work to perfect the theory of elasticity, which was crucial in the construction of the Eiffel Tower later in the century. (Nevertheless, the name of Sophie Germain is *not* inscribed on the base of the Eiffel Tower among the seventy-two scholars whose work made possible its construction!)

The problem of understanding electromagnetic phenomena fascinated many mathematicians, including Bernard Riemann. Riemann's genius led him to make fundamental contributions to all the areas in which he worked. From the geometrical discoveries mentioned above to his work in electromagnetism, all he did was inspired. Sometimes he was so inspired that he left out many of the details, leaving the work of completing his proofs to his successors. In his applied work, Riemann was perfectly willing to use arguments from physics when he did not have mathematical proofs. This, too, left interesting questions for his successors to unravel. In one of his papers, he says something like "it is reasonable to believe that..." The assertion that Riemann thought was reasonable is now known as the *Riemann Hypothesis*, and it is still to be proved.

Near the end of the 19th century, Felix Klein showed that the new non-Euclidean geometries and the new algebraic theories were linked. One could understand the different geometries by analyzing the algebra of the transformations allowed by them. This demonstrated once again that rigor and abstraction were the keys to progress in mathematics. It also showed that this point of view allowed one to unify disparate branches of mathematics. In a discipline that was growing by leaps and bounds, it was very important to find unifying ideas that would allow people to understand wider ranges of mathematics.

The trend toward unification of mathematics was personified by Henri Poincaré. His almost superhuman memory and powers of logical apprehension enabled him to produce valuable contributions to number theory, algebra, geometry, analysis, astronomy, and mathematical physics. He wrote philosophical works on science and mathematics. Among many other things, Poincaré's work led to completely new points of view in dynamics, particularly in the dynamics of the solar system. It was in this context that Poincaré encountered the first examples of what we now call "deterministic chaos."

As the century came to an end, mathematics was becoming more and more a profession. Most mathematicians worked at universities, where they both taught and did research. Research seminars became a normal feature of university life, and the Ph.D. became the standard entry point to an academic career. The lines separating pure mathematics, applied mathematics, and theoretical physics became more distinct. Professional societies and journals were created. Congresses and meetings began to happen regularly. In 1897, the first International Congress of Mathematicians was held in Zürich. The next was in Paris in 1900, and from then on an ICM has been held every four years (with interruptions due to war), bringing together mathematicians from all over the world.

The mathematics of the 19th century is quite technical, and so are most historical studies of the period. The best sources for the overall story are [99] and [76].

Abstraction, Computers, and New Applications

The end of the 19th century was marked by an important event. In 1900, the organizers of the International Congress of Mathematicians invited David Hilbert, then the most prominent German mathematician, to give a lecture. Hilbert's address focused on the role of unsolved problems in driving mathematical research. He outlined 23 such problems that he felt would be important in the new century. For the most part, he guessed right. The investigation of these problems led, directly or indirectly, to many important advances in mathematics. The solution of even part of one of Hilbert's problems carried with it international recognition for the solver.

Even Hilbert, however, could not foresee how mathematics would grow in the 20th century. There were more and more mathematicians, more and more journals, more and more professional societies. The phenomenal growth that had begun in the 1800s continued, with mathematical knowledge doubling every twenty years or so. More original mathematics has been produced *after* astronauts first walked on the moon than there had been in all previous history. In fact, it is estimated that 95% of the mathematics known today has been produced since 1900. Hundreds of periodicals published all over the world devote a major share of their space to mathematics. Each year the abstracting database *MathSciNet* publishes many thousand synopses of recent articles containing new results. The 20th century (and this new century so far) is justifiably called the "golden age of mathematics."

Quantity alone, however, is not the key to the unique position the current era occupies in mathematical history. Beneath this astounding proliferation of knowledge, there is a fundamental trend toward unity. The basis for this unity is abstraction. This has led in two directions. On the one hand, new, more abstract subfields of mathematics have emerged to become established areas of research in their own right. On the other, researchers working on truly big classical problems, such as Fermat's Last Theorem or Hilbert's 23 problems, have become increasingly adept at using new techniques from one area of mathematics to answer old questions in another. As a result, the 20th century was a time when many old questions were finally answered and many new questions came to the fore.

The increase in the sheer amount of mathematics and in the level of abstraction has led, inevitably, to a growth in specialization. Most mathematicians today know their research areas well, the surrounding topics less well, and distant parts of the subject very little, if at all. The unifying power of the abstract point of view makes up for this somewhat, but only the very best mathematicians have a full view of the field. Given all that, we can provide no more than a brief and very incomplete description of 20th century mathematics. We have chosen to highlight the debate on the foundations of mathematics, the role of abstraction, the invention of the computer, and the wide range of new applications.

In the first few decades of the 20th century, mathematicians and philosophers investigated the foundations of the subject. What *is* a number? Is mathematical knowledge certain? Can we *prove* that it is impossible for mathematics to contradict itself? There was intense debate about these questions. On one side were the "formalists," whose program was to show that philosophical qualms about mathematical ideas could be laid to rest by a study of formal manipulation of symbols. The hope was to be able to justify working with infinite sets and other contested ideas by giving a proof, using nothing but finite methods, that such ideas would never lead to a contradiction. On the other side were the "intuitionists," who argued that many mathematical ideas were in fact *not* well-founded. They wanted to reform the whole subject by eliminating most appeals to infinite sets and getting rid of the "law of the excluded middle" (the Aristotelian principle that said that if one can prove that not-A is false, then A must be true).

Neither current got very far. The intuitionists' proposals were too radical for most mathematicians. The formalists' ideas were closer to being acceptable, but their program lost much of its attractiveness after the 1930s. What happened was that Kurt Gödel found a way to *prove* that it was *impossible* to find a proof that contradictions would not arise. Gödel's work established for the first time that some things could not be proved.

This had a curious effect. For mathematicians working on foundations, it was a real blow, one that had to be absorbed somehow. For other mathematicians, however, it basically meant that the work on foundations would not be able to offer them much help in solving the big problems. So they went on with the work of trying to prove theorems and solve problems.

As the foundational debate happened, the move to abstraction became the dominant theme of mathematics early in the century. It was not just that it was fashionable (though that surely was part of it). It was also that the abstract method was *powerful.* Using it, old problems were either disposed of or cast in a new light. Soon mathematics was dominated by abstract analysis, topology, measure theory, functional analysis, and other such areas. In essence, these were vast generalizations of material developed in the 19th century, taking an abstract point of view. The new point of view revealed which ideas were important and which were not. As a result, it often led to new discoveries.

Nowhere was the change more visible than in algebra. The subject became far more general than it had ever been before. Galois's insight that one must consider whole classes of algebraic operations in one blow had been developed further by Dedekind in the second half of the 19th century. In the 1920s, these ideas were picked up and extended by Emmy Noether and Emil Artin. In their hands, the structure and language of abstract algebra became a powerful tool.

Late in the 1930s, a group of young French mathematicians came together with the intention of revolutionizing the subject. They felt that the new ideas had not been sufficiently internalized by the mathematical community, especially in France. It was time to bring down the "old guard." Their plan was to do two things. First, they would collectively write a multivolume textbook, a compendium of all of fundamental mathematics. With a nod towards Euclid, they called their work the *Elements of Mathematics*. Since they were writing it collectively, they adopted a pseudonym: the author of the new *Elements* was "Nicolas Bourbaki."[26]

The members of the Bourbaki group had great fun with their collective persona. They created a biography for Nicolas Bourbaki, gave him a university affiliation, and even, at one point, sent out invitations to his daughter's wedding. When the *Encyclopaedia Britannica* said that Bourbaki was a collective pseudonym and that therefore no such person existed, they received an angry letter from Nicolas Bourbaki "himself" questioning the existence of the author of the encyclopedia entry! The founders of Bourbaki believed that creative mathematics is primarily a young people's sport, so they agreed to retire from the society before the age of fifty and elect younger colleagues to replace them.[27] Thus, Nicolas Bourbaki has become a renowned but mysterious international scholar, always at the peak of his professional productivity, supplying the scientific world with a continuing series of modern, clear, accurate expositions in all fields of contemporary mathematics.

There was a second part to the strategy of the Bourbaki group.

---

[26]It has been suggested that the name "Bourbaki" was inspired by a statue in Nancy, France, of General Charles D. S. Bourbaki from the Franco-Prussian War. The motivation for "Nicolas" is unclear, but its choice makes the resulting initials, N. B., an attractive bonus.

[27]It is rumored that some of the original founders regretted having made this rule when it came time for them to leave.

They decided to hold a regular seminar about what was going on in mathematics; this became known as the "Séminaire Bourbaki." Now held in Paris three times a year, it is still one of the most important seminars on recent advances in the field, in which top scholars discuss important new ideas and theorems to a packed audience of mathematicians from all over the world.

Bourbaki's influence was mostly felt through the *Elements*. Since these took many years to write, their impact came around mid-century or later. The books take an uncompromisingly abstract point of view. Nevertheless, they give a precise and reliable account of each of the fields they cover. Many have blamed Bourbaki for having led mathematics teachers to adopt a formal and abstract approach. But the truth is that the *Elements of Mathematics* was never meant to be a model of mathematical pedagogy. Rather, these volumes were meant to bring together, formalize, and make precise large chunks of mathematics, and this they did (mostly) successfully.

By the last decades of the century, Bourbaki seemed to have lost some of "his" original energy. The *Séminaire* was (and is) still going strong, but the rate of publication of the *Elements* slowed to a crawl. In part this was due to the success of Bourbaki's program, but it also reflected a change in the mood of mathematicians, a move away from abstraction and towards concrete examples, computer experimentation, and applications.

The second half of the 20th century was marked by the invention and development of computers. (See Sketches 23 and 24.) At the beginning, mathematicians were deeply involved in this process. They analyzed the possibility of the new machines, invented the notion of a "computer program," helped build the first devices, and provided difficult computational problems to test them. Soon, however, computer science went its own way and focused on its own problems.

It is as tools that computers have had the greatest impact on mathematics. Computers have changed mathematics in at least three ways. First, they allowed mathematicians to test conjectures and discover new results. This allows for "experimental mathematics." Suppose one is trying to find out whether there are infinitely many primes of the form $n^2 + 1$, a problem that is still unsolved. One can test whether this is likely or not by using a computer to plug in several million values of n and testing the resulting number for primality. If we find many primes, we may be persuaded that in fact there will be infinitely many of them. Notice that this doesn't decide the issue. The hard work of finding a proof remains to be done. But it can offer clues and it can make us more or less confident that what we are trying to prove is indeed true.

The second change has to do with simulation and visualization. Computers can make pictures of data, pictures that are often far more illuminating than the numerical data by itself. Furthermore, they allow us to use numerical computation to find approximate solutions to equations. Thus we can use them to understand situations whose precise description is too complicated for complete mathematical analysis. This has, of course, revolutionized applied mathematics. Nowadays, approximate solutions to complicated differential equations are a central part of that endeavor. But it has also impacted pure mathematics. A good example of this effect is the discovery of fractals. These highly complex structures had been considered before, early in the 20th century, but at that point they just seemed horrendously complicated and impossible to deal with. Once we learned to use computers to draw their pictures, however, we discovered that fractals can be beautiful, and a whole new field of mathematics was born.

The third change is related to what are known as *computer algebra systems*. These are computer programs that can "do algebra." They can work with polynomials, trigonometric functions, exponentials, and lots more. They can add, multiply, factor, take derivatives and integrals, and compute series approximations. In other words, they can do much of what used to be the computational content of school and  college mathematics. As these systems become more widely available, there is less and less reason to teach students to do elaborate computations by hand. As a result, mathematicians need to rethink what should be taught and how.

The 20th century also saw a great widening in the range of applications of mathematics. In the late 19th century, "applied mathematics" was almost synonymous with "mathematical physics." This quickly changed. Probably the first change was the development of statistics, initially with a view to applications in biology. The value of statistics as a tool for analyzing data was soon evident to scholars everywhere, and other applications quickly followed. (See Sketch 22 for the early history of statistics.)

More and more mathematical ideas were found to be useful. Mathematical physics began to use probability and statistics, Riemannian geometry, Hilbert spaces, and group theory. Chemists discovered that crystallography had a large mathematical component. Topological ideas proved relevant to the study of the shape of molecules. Biologists used differential equations to model the spread of disease and the growth of animal populations.

Beginning in World War I, governments discovered that thinking mathematically about practical problems led to useful results, and "operations research" was born. In World War II, mathematicians had a crucial role in many ways, most famously in developing the science of cryptography and breaking the German Enigma code. Telephone networks and, later, the Internet were studied mathematically. "Linear programming" was developed to deal with finding the "best way" to do things in all sorts of areas, from industry to government to the military. Computer modeling allowed all sorts of new applications to biology, from the dynamics of populations to studies of blood circulation, neurons, and how animals move. Finally, towards the end of the century, mathematics and physics got together again to produce new and sophisticated physical theories.

Perhaps the most interesting account of the transformation of mathematics after the late 19th century is [77], which sees it in the context of "modernism" in the arts. Another way to get a feel for the early twentieth century is to read about the solvers of the Hilbert problems in [178]. For more recent work, see also [4], [3] and [5], which contain interesting profiles of living mathematicians.

Mathematics Today

The last few decades have continued to be a "golden age" for mathematics. Many old problems have been solved and many new ideas have been introduced. Both pure and the applied mathematics have been strikingly successful. In this concluding section we mention a few of these new discoveries.

At the dawn of the 20th century, Hilbert presented a list of problems he deemed important. Those problems were an important motivation for mathematicians in the first half of that century. As the 21st century began, the Clay Mathematics Institute decided to follow his lead. They chose seven mathematical problems and offered a one million dollar prize for the solution of each of them. The seven "millennium problems" cover the full spectrum from the purest of pure mathematics to mathematical issues related to fluid flow, particle physics, and the theory of computation.

One of the millennium problems went back all the way to Poincaré's work in the late 19th century. It concerned the geometry of three-dimensional space. In the 20th century, it was generalized to higher dimensions, and it turned out to be easier to solve in that case. In dimensions five and higher, the conjecture was proved by Stephen Smale in the 1960s. In dimension four, it was proved by Donald Sullivan in the 1980s. But the original three-dimensional problem remained unassailable until 2002, when Grigory Perelman found a proof. Remarkably, he refused to accept the million dollar prize!

One interesting development has to do with an old conjecture due to Kepler concerning how to place the largest number of spheres into the smallest possible space. Imagine you have a bunch of equal-size marbles; what arrangement of the marbles will pack them most tightly? Kepler came up with a guess, but no one was able to prove it until 1998... maybe! Thomas Hales gave a proof, but his proof requires a very large number of cases to be checked by computer. He sent it to the *Annals of Mathematics*, which asked 12 experts to check the proof. They ended up deciding, in 2003, that they were "99% sure" that it was correct, but that they could not be certain because they had no way to check the computer part. Is 99% certainty enough?

Number theory was another growth area. As we explain in Sketch 13, Fermat's Last Theorem was proved in 1994. But there have been many other impressive results. In 2004, Ben Green and Terry Tao proved that one can find arbitrarily long arithmetic progressions made up entirely of prime numbers, such as $(3, 7, 11)$, with step 4 and the longest one known, starting with $43, 142, 746, 595, 714, 191$ followed by 25 other primes separated by a step of $5, 283, 234, 035, 979, 900$. A related problem also saw a huge advance: mathematicians have long suspected that there are infinitely many prime numbers that are two units apart, such as 11 and 13 or 29 and 31, but no one knows how to prove that. But in 2013, Yitang Zhang surprised the world by proving that if we allow a bigger fixed gap instead of two, then there are infinitely many such primes. With the work of many others, we now know that there are infinitely many pairs of primes with a gap no larger than 246.

Number theory is often thought of as the purest of pure mathematics, but in the 1970s we learned otherwise. You may have noticed that multiplying is easier than factoring: if someone asks you to compute 463 times 2029, you can do it easily, but if they hand you $939, 427$ and ask you to factor it, that's much harder. In the 1970s, Ron Rivest, Adi Shamir, and Leonard Adleman figured out a way to use this to construct a way of sending secret messages without having to meet the other person to exchange a secret key. We all use this idea today: it is what allows us to encrypt information sent over the Internet. The neat thing for mathematics, however, is that suddenly a better way to factor numbers becomes a powerful tool! A lot of effort has gone into that. Some good methods have been found, but none that is good enough to break the new encryption method.

Some of the most exciting developments are those connected to other fields. In mathematical physics, String Theory has brought to light new and deep mathematical questions. Mathematical and computational biology is beginning to offer up significant insights. Probabilistic methods now pervade mathematical modeling and engineers have been using more and more powerful mathematical techniques. The invention of wavelets has had a deep effect on the processing of signals and images.

We see even more diversity if we broaden our view to include those users and producers of mathematics that are outside of the usual academic institutions. Many of the most interesting and potentially useful issues confronting mathematics today are "crossover questions" involving connections between mathematics and science. In the world of finance, mathematics has become so important that some people have looked for mathematical causes underlying the financial crisis of 2008. Technological advances in many areas of business and commerce are increasingly dependent on more sophisticated mathematical ideas. Mathematics lies at or near the heart of innovations in many of today's multimillion-dollar industries. Aircraft design, genetic research, missile defense systems, CD players, epidemic control, GPS satellites and space stations, cellular phone networks, marketing and political surveys, personal computers and calculators, "morphing" and other special visual effects used in movies and video games, computer animation, and the electronic hardware and software tools that handle the everyday affairs of almost every business of any size — all of these things and many, many others depend on mathematical ideas and require mathematical experts to implement.

Mathematics today involves a huge number of people doing many different things. At universities and research institutes, high-powered research continues to push at the boundaries of our knowledge. Outside academia, many people use and develop mathematical techniques every day. While many of these people don't describe themselves as "mathematicians," their activities help push mathematics and its applications forward. And all of these people are supported by a vast network of teachers and educators at all levels.

So much work in so many different areas gives the impression of fragmentation, but mathematics today, as seen "from the inside," is both more diverse and more unified than it has ever been. It is more abstract and yet has a broader applicability to all areas of modern life than ever before. It is true that research mathematicians are usually specialists in one area or another, and their research almost always lies far beyond the reach of interested amateurs, even those with college degrees in mathematics. Any new theorem in today's research journals is likely to be incomprehensible to most mathematicians whose specialty does not include its topic. But despite the diversity, there is underlying unity. Mathematics has always been about big ideas, and these are still at the center of the field, though they may not be apparent at first look.

The view "from the outside" is understandably a bit confusing. On the one hand, mathematics is seen as an esoteric, daunting subject about which even many well-educated people unblushingly admit ignorance. On the other hand, it is seen as an essential part of modern prosperity, security, and comfort, so that the mathematically adept are pursued as valuable human resources. Governments keep trying to get more people to learn mathematics, but the subject remains difficult and inaccessible to far too many students.

This ambivalence has created a dilemma for mathematical educators. Do they allow the internal demands of the subject to dictate ever more rigorous training in the traditional areas of algebra, geometry, calculus, and the like, so that the next generation of experts will have a firm foundation for their research? Or do they let the external needs of society dictate a broader, less intensive education in mathematical ideas that will allow everyone to become mathematically literate citizens who can interact intelligently with the experts? This apparent conflict of goals has created tension and turmoil in mathematics education today, but the hopeful view is that it is creative, constructive turmoil as educators find ways to achieve both. It is critically important that they do so; tomorrow's world will not accept less.

Keeping Count
Writing Whole Numbers

The problem of how to write numbers efficiently has been with humanity for as long as there have been sheep to count or things to trade. The simplest, most primitive way to do this was (and still is) by tallying — making a single mark, usually | or something just as simple, for each thing counted. Thus, *one, two, three, four, five,...* were written or carved as

| || ||| |||| |||||

and so on. People still use this for scorekeeping in simple games, class elections, and the like, sometimes bunching the stroke marks by fives.

The simplicity of the tally system is its greatest weakness. It uses only one symbol, so *very* long strings of that symbol are needed for writing even moderately large numbers. As civilization progressed, various cultures improved on this method by inventing more number symbols and combining them in different ways to represent larger and larger numbers.
During the past six millennia, more than 100 distinct numeration systems have been used by various groups of people at various times.[1] Examining a few of those earlier numeration systems can illustrate the power and convenience of our current system for writing numbers.

Sometime before 3000 B.C., the ancient Egyptians improved on the tally system by choosing a few more number symbols and stringing them together until their values added up to the number they wanted. These symbols were "hieroglyphic"; that is, they were small picture drawings of common things. The basic pictures of the Egyptian system and their numerical values appear in Display 1.

In this system the number *one hundred thirteen*, for instance, could be written as ℚ ∩ | | | or as | ∩ | ℚ | or as | | ℚ | ∩. The order of the symbols didn't matter, as long as they added up to the right value. Of course, the pictures with the largest possible values in each case were used to make the writing process more efficient. Even so, larger numbers often required fairly long strings of symbols. For instance, the number they would write as 1,213,546 is

𓂽 𓆐 𓆐 𓍢 𓏤𓏤𓏤 ℚ ℚ ℚ ℚ ℚ ∩∩∩∩|||||

---

[1] See [29].

Symbol	Interpretation	Number represented
|	stroke	1
∩	heel bone	10
℮	coiled rope	100
𐃺	lotus flower	1000
𓂭	pointed finger	10,000
⌇	tadpole	100,000
𓁨	astonished man	1,000,000

Egyptian hieroglyphic numerals

Display 1

One of the earliest artifacts showing this system in use is a royal Egyptian mace dating from about 3000 B.C., now in a museum at Oxford, England. It is a record of a successful military campaign, adorned with numbers in the hundreds of thousands and even in the millions.

Much of our knowledge of Egyptian hieroglyphic numerals comes to us by way of inscriptions carved on monuments and other durable artifacts. The hieroglyphic symbols were used for such purposes throughout the entire span of the ancient Egyptian civilization, even up to the 5th century A.D. Around 2600 B.C., the scribes of Egypt developed a more efficient system for writing numerals with ink on papyrus. The new system, called "hieratic," used distinct basic symbols for each unit value from 1 to 9 and for each power-of-ten multiple from 10 to 9000. This allowed for more compact expressions, at the expense of an added strain on the writer's and reader's memories.

The region known as Mesopotamia (now part of Iraq), has been called the "cradle of civilization." At least ten distinct numeration systems were used at one time or another in that region during the three millennia after about 3500 B.C. From the period between 2000 and 1600 B.C. comes the one of particular interest to us, a system that the Babylonian scribes used in their computations. It was based on two wedge-shaped ("cuneiform") symbols, represented here as 𒁹 and 𒌋 . Those basic symbols were quickly and easily formed in soft clay tablets with a simple scribing tool. When baked hard, these tablets formed a permanent record, and many of them have survived.

This was a positional or "place-value") system; that is, it used the position of the symbols to determine the value of a symbol combination. The scribes multiplied successive groups of symbols by increasing powers of sixty, much as we multiply successive digits by increasing powers of ten. Thus, their system is called a *sexagesimal* system, just as ours is a *decimal* system. The numbers 1 to 59 were represented by combinations of the two basic symbols used additively, with each 〖 representing *one* and each 〈 representing *ten*. For instance, *twenty-three* was written as 〈 〈 〖〖〖 .

The numbers from 60 to 3599 were represented by two groups of symbols, the second group placed to the left of the first one and separated from it by a space. The value of the whole thing was found by adding the values of the symbols within each group, then multiplying the value of the left group by 60 and adding that to the value of the right group. For instance,

A hand tablet[2]

$$\langle \mbox{〖〖} \quad \langle\langle\langle\mbox{〖}$$

represented

$$(10 + 1 + 1) \cdot 60 + (10 + 10 + 10 + 1) = 12 \cdot 60 + 31 = 751.$$

They wrote numbers 3600 $(= 60^2)$ or greater by using more combinations of the two basic wedge shapes, placed further to the left, each separated from the others by spaces. Each single combination value was multiplied by an appropriate power of 60 — the combination on the far right by 60^0 $(= 1)$, the second from the right by 60^1, the third from the right by 60^2, and so on. For example, 7883 was thought of as $2 \cdot 3600 + 11 \cdot 60 + 23$ and written

$$\mbox{〖〖} \quad \langle\mbox{〖} \quad \langle\langle\mbox{〖〖〖} .$$

A major difficulty with the Babylonian system is the ambiguity of the spacing between symbol groups. For instance, it is not clear how 〖   〈   should be interpreted; it could be

$$1 \cdot 60^2 + 10 \quad \text{or} \quad 1 \cdot 60^3 + 10 \cdot 60^2 \quad \text{or} \quad 1 \cdot 60 + 10 \quad \text{or} \quad \ldots$$

By 400 B.C. or so, the Maya civilization of Central America had a numeration system similar to that of the Babylonians, but free from this spacing difficulty. They had two basic symbols, a dot "·" for the number one, and a bar "——" for the number five. The numbers one through nineteen were written like this:

The Maya used groupings of the basic symbols to represent larger

---

[2]This is a drawing of Ur Excavation Tablet 236 (reverse) by Eleanor Robson; see [146], figure A.5.10. Used by permission.

numbers, often arranged vertically. They were evaluated by adding the place-value amounts for each group. The lowest group represented single units; the value of the second group was multiplied by 20, the value of the third by $18 \cdot 20$, the value of the fourth by $18 \cdot 20^2$, the value of the fifth by $18 \cdot 20^3$, and so on. This Mayan positional system, which was based on twenty except in the third place, apparently was used only for recording dates in their Long Count calendar. Thus, the peculiar use of $18 \cdot 20 = 360$ as the third place value probably stems from its approximation of the number of days in a year.

The spacing difficulty of the Babylonian system was circumvented by the invention of a shell-like symbol, ⬭ , to show when a grouping position was skipped. For example, 52,572 was written as follows:

	$(5 + 1 + 1) \cdot (18 \cdot 20^2)$	=	50,400
	$(5 + 1) \cdot (18 \cdot 20)$	=	2,160
	$0 \cdot 20$	=	0
	$5 + 5 + 1 + 1$	=	+ 12
			52,572

The Mayan symbol was better than the Babylonians' ambiguous spacing. However, since their culture was not known to Europeans until many centuries later, their system had no influence on the development of Western numeration. The roots of Western European culture go back mainly to the ancient Greeks and Romans. In many ways, both the Greek and the Roman systems were more primitive than the relatively efficient Babylonian system. The main Greek numeration system required the 25 letters of their alphabet and two extra symbols — nine for the units, nine for the tens, and nine for the hundreds — and a special mark for representing numbers larger than 1000.

The Roman Empire's domination of civilized Europe from about the first century B.C. to the fifth century A.D. made Roman numeration the commonly accepted European way of writing numbers for many centuries afterwards, even into the Renaissance. Like the Egyptian system, Roman numeration is additive and not positional (with one minor exception). Display 2 shows its basic symbols and their corresponding values. The values of these basic symbols were added to determine the value of the entire numeral. For instance,

CLXXII $= 100 + 50 + 10 + 10 + 1 + 1 = 172.$

Symbol	Value
I	1
V	5
X	10
L	50
C	100
D	500
M	1000

Roman numerals

Display 2

Larger numbers were written by putting a bar over a set of symbols to indicate multiplication by 1000. Thus, $\overline{V} = 5000$ and

$$\overline{VII}CLXV = 7000 + 100 + 50 + 10 + 5 = 7165.$$

A peculiar feature of the Roman system is a subtractive device, introduced in later years for efficiency. The values of some symbol combinations are found by subtraction: If a basic symbol in a numeral has a smaller value than the one immediately to its right, then the smaller value is subtracted from the larger one to get the value of the pair. For instance,

$$IV = 5 - 1 = 4.$$

Ambiguity is avoided by requiring that only symbols representing powers of ten may be subtracted, and they may only be paired with the next two larger values:

I may be paired with V and X, but not with L, C, D, or M.

X may be paired with L and C, but not with D or M.

C may be paired only with D and M.

For instance,

$$MCMXCIV = 1000 + 900 + 90 + 4 = 1994.$$

By this method, no more than three adjacent copies of the same basic symbol are needed in any numeral.

Our current method for writing numbers is called the Hindu-Arabic system. Invented in India sometime before 600 A.D. and refined over succeeding centuries, it was picked up by the Arabs during the Islamic expansion into India in the 7th and 8th centuries. The Europeans, in turn, learned it from the Arabs. It uses place value and is based on powers of ten. Its basic symbols — 0, 1, 2, 3, 4, 5, 6, 7, 8, 9 — are called *digits* and represent the numbers zero through nine.

Nobody quite knows why the number ten was originally chosen as the base for this system. The standard conjecture is that it was more biological than logical. Research indicates that this numeration system, like many others, emerged from finger counting, so it was natural that the base number should correspond to the number of fingers we human beings have. The very word we use for the basic numerals reflects this fact; *digitus* is the Latin word for *finger*.

Despite its simplicity and efficiency, the Hindu-Arabic method of writing numbers did not displace the use of the Roman numeration

system in Europe for several centuries. Old habits die hard. There were
also practical reasons. For example, people worried about how easily a
"2" can be changed into a "20" in the Hindu-Arabic system. Because
of this, laws were passed saying that in legal documents numbers had
to be written out in words. We still do this when we write checks.

The Roman numeration system is not very good for computation.
(Try multiplying MCMXLVII by CDXXXIV without translating into
Hindu-Arabic numerals; then try it again in our system.) When it
was in use, computation was done on a counting board or an abacus,
rather than on paper. One of the changes brought on by the Hindu-
Arabic system is the fact that it is possible to compute directly with
the numbers as written. The availability of cheap paper helped the
new numbers to catch on. The advent of calculators has in some ways
brought us back to having two systems: one for recording numbers and
one, electronic, for actually doing computations. Of course, it's not
impossible to compute with Roman numerals. It's just complicated.
Logician Martin Davis tells[3] that

> In 1953, I had a summer job at Bell Labs in New Jersey,
> (now Lucent), and my supervisor was Claude Shannon [a
> computer pioneer and the creator of the mathematical the-
> ory of communication]. On his desk was a mechanical cal-
> culator that worked with Roman numerals. Shannon had
> designed it and had it built in the little shop Bell Labs had
> put at his disposal. On a name plate, one could read that
> the machine was to be called "Throback I."

Though we still use Roman numerals for ornamental purposes, there
is no chance we'll ever abandon the compact, convenient, and useful
Hindu-Arabic system. The power of the Hindu-Arabic system stems
from its efficient positional structure, which is based on powers of ten.
That's why we call it a *decimal place-value system*.

For a Closer Look: For more on writing numbers, see Sketches 3
and 4. There are more detailed discussions in standard histories and
also in [145] and [19]. For book-length treatments, see [123] or [29].
Denise Schmandt-Besserat, who has done important scholarly work in
this area, has written [154], a children's picture book on counting and
numeration. It is easy to read and well illustrated.

---

[3]In a posting on the *Historia Mathematica* electronic discussion group.

2 Reading and Writing Arithmetic
The Basic Symbols

How would you write the statement "When 7 is subtracted from the sum of 5 and 6, the result is 4" using arithmetic symbols? Would you write $(5+6) - 7 = 4$? Probably. If you did, your expression would have several advantages over the sentence it represents: it's more efficient to write, clearer and less ambiguous to read, and understandable by almost anyone who has studied elementary arithmetic, regardless of the country they're in or the language they speak.

The symbols of arithmetic have become universal. They are far more commonly understood than the letters of any alphabet or the abbreviations of any language. But that hasn't always been the case. The ancient Greeks and their Arab successors didn't use any symbols for arithmetic operations or relations; they wrote out their problems and solutions in words. In fact, arithmetic and algebra statements were written only in words by most people for many, many centuries, right through the Middle Ages.

Arithmetic symbols arose as written shorthand in the early years of the Renaissance, with very little consistency from person to person or from country to country. With the invention of movable-type printing in the 15th century, printed books began to exhibit a little more consistency. Nevertheless, it was a long time before today's symbols became a common part of written arithmetic.

Here are some ways in which $(5 + 6) - 7 = 4$ would have appeared during the centuries from the Renaissance to now. In most cases the date given is the year that a particular book was published; think of it as an approximation of the time in which that notation was in fairly common use, at least by mathematicians of a particular region.

1470s: Regiomontanus in Germany would have written

$$5 \; et \; 6 \; \widehat{\imath \varphi} \; 7 \text{——} 4$$

(The word *et* is "and" in Latin.)

1494: In Luca Pacioli's *Summa de Aritmetica*, widely used in Italy and other parts of Europe, this would have appeared as

$$5 \; \tilde{p} \; 6 \; \tilde{m} \; 7 \text{——} 4$$

The grouping of the first sum probably would have been ignored, assuming that it obviously was to be done first. This notation for plus and minus became very common throughout much of Europe.

1489: About the same time in Germany, our now-familiar plus and minus signs appeared in print for the first time, in a commercial arithmetic book by Johann Widman. Widman had no symbol for equality, so his version of our statement would likely have been something like

$$5 + 6 - 7 \text{ das ist } 4.$$

(The German phrase "das ist" means "that is.") But Widman also used + as an abbreviation for "and" in a non-numerical sense and − as a general mark of separation. The idea that these symbols had primary mathematical meanings was not yet clear.

1557: The first use of + and − in an English book occurs in Robert Recorde's algebra text, *The Whetstone of Witte*. In this book Recorde also introduces === as a symbol for equality. He justifies it by saying, "no two things can be more equal" (than parallel lines of the same length). His other signs are elongated, too. He might have written

$$5 +\!\!-\, 6 -\!\!-\!\!- 7 =\!\!= 4$$

Recorde's notation was not immediately popular. Many European writers preferred to use $\tilde{p}$ and $\tilde{m}$ for plus and minus, particularly in Italy, France, and Spain. His equality sign didn't appear in print again for more than half a century. Meanwhile, the symbol = was being used for other things by some influential writers. For instance, in a 1646 edition of François Viète's collected works, it was used for subtraction between two algebraic quantities when it was not clear which was larger.[1]

1629: Albert Girard of France would have represented the left side of our equation either as $(5 + 6) - 7$ or as $(5 + 6) \div 7$; to him they meant the same thing! In fact, $\div$ was widely used for subtraction during the 17th and 18th centuries, and even into the 19th century, particularly in Germany.

1631: In England, William Oughtred published a highly influential book, known as *Clavis Mathematicae*, emphasizing the importance of using mathematical symbols. His use of +, −, and = for addition, subtraction, and equality contributed to the eventual adoption of these symbols as standard notation. However, if Oughtred wanted to emphasize the grouping of the first two terms of our expression, he would have used colons. Thus, he might have written

$$: 5 + 6 : -7 = 4$$

In this same year, Recorde's long equality symbol appeared in an influential book by Thomas Harriot, along with > and < for "greater than" and "less than," respectively.

---

[1] In other words, it was used for the absolute value of the difference.

1637: René Descartes's *La Géométrie*, the book that simplified and regularized much of the algebraic notation we use today, was also responsible for delaying universal acceptance of the $=$ sign for equality. In this and some of his later writings, Descartes used a strange symbol for equality.[2] In this book Descartes also used a broken dash (a double hyphen) for subtraction, so his version of our equation would have been

$$5 + 6 -- 7 \; \infty \; 4$$

Descartes's algebraic notation spread rapidly through the European mathematical community, often carrying with it the strange new symbol for equality. Its use persisted in some places, particularly in France and Holland, into the early 18th century.

Early 1700s: Parentheses gradually replaced other grouping notations, thanks largely to the influential writings of Leibniz, the Bernoullis, and Euler. Thus, by the time the American colonies were preparing to separate from British rule, the most common way to write our simple equation was the one we use today:

$$(5 + 6) - 7 = 4.$$

Does it surprise you that there were so many different ways of symbolizing addition, subtraction, and equality? Do you find it strange that people should put up with such ambiguity? It's not really so different from some things we routinely do nowadays. For instance, we still have at least four different notations for multiplication:

- $3(4 + 5)$ means 3 times $(4 + 5)$. Writing multiplication as juxtaposition (just placing the quantities to be multiplied side by side) dates back to Indian manuscripts of the 9th and 10th centuries and to some 15th century European manuscripts.

- The $\times$ symbol for multiplication first appears in European texts in the first half of the 17th century, most notably in Oughtred's *Clavis Mathematicae*. A larger version of this symbol also appears in *Géométrie*, a famous textbook by Legendre published in 1794.

- In 1698, Leibniz, bothered by the possible confusion between $\times$ and x, introduced the raised dot as an alternative sign for multiplication. It came into general use in Europe in the 18th century and is still a common way to symbolize multiplication. Even today, $2 \cdot 6$ means 2 times 6.

- Today, calculators and computers use an asterisk for multiplication; 2 times 6 is entered as $2 * 6$. This very modern notation was

---

[2]Cajori ([22], p. 301) is of the opinion that it is the astronomical sign for Taurus rotated a quarter turn to the left.

used briefly in Germany in the 17th century;[3] then it disappeared from arithmetic until this electronic age.

Signs for division are just as diverse. We write 5 divided by 8 as $5 \div 8$, or $5/8$, or $\frac{5}{8}$, or even as the ratio $5{:}8$. The use of $\div$ for division, rather than subtraction, is primarily due to a 17th century Swiss algebra book, *Teutsche Algebra* by Johann Rahn. This book was not very popular in continental Europe, but a 1668 English translation of it was well received in England. Some prominent British mathematicians began using its notation for division. In this way, the $\div$ symbol became the preferred division notation in Great Britain, in the United States, and in other countries where English was the dominant language, but not in most of Europe. European writers generally followed the lead of Leibniz who (in 1684) adopted the colon for division. This regional difference persisted into the 20th century. In 1923, the Mathematical Association of America recommended that both signs be dropped from mathematical writings in favor of fractional notation, but that recommendation did little to eliminate either one from arithmetic. We still write expressions such as $6 \div 2 = 3$ and $3 : 4 :: 6 : 8$. (For the history of fractional notation, see Sketch 4.)

In this brief sketch we have skipped over many, many symbols that were used from time to time in written or printed arithmetic but are now virtually (and happily) forgotten. Clear, unambiguous notation has long been recognized as a valuable ingredient in the progress of mathematical ideas. In the words of William Oughtred in 1647, the symbolical presentation of mathematics "neither racks the memory with multiplicity of words, nor charges the phantasie [imagination] with comparing and laying things together; but plainly presents to the eye the whole course and process of every operation and argumentation."[4]

For a Closer Look: For a readable history of mathematical notation that also reflects on its power and importance, see [122]. The information for this sketch was drawn primarily from [22], which remains a good reference. See also *Earliest Uses of Various Mathematical Symbols*, a website maintained by Jeff Miller.

---

[3]See [22], p. 266.

[4]From William Oughtred's *The Key of the Mathematicks* (London, 1647), as quoted in [22], p. 199, adjusted for modern spelling.

3

Nothing Becomes a Number
The Story of Zero

Most people think of zero as "nothing." The fact that it is *not* nothing lies at the root of at least two (some would say three) important advances in mathematics. The story begins in Mesopotamia, the "cradle of civilization," sometime before 1600 B.C. By then, the Babylonians had a well developed place-value system for writing numbers. It was based on grouping by sixty, much as we count 60 seconds in a minute and 60 minutes (3600 seconds) in an hour. They had two basic wedge-shaped symbols — ∇ for "one" and $\triangleleft$ for "ten" — which were repeated in combination to stand for any counting number from 1 to 59. For instance, they wrote 72 as

$$\nabla \quad \triangleleft \nabla \nabla$$

with a small space separating the 60s place from the 1s place.[1]

But there was a problem with this system. The number 3612 was written

$$\nabla \qquad \triangleleft \nabla \nabla$$

(one $3600 = 60^2$ and twelve 1s) with a little extra space to show that the 60s place was empty. Since these marks were made quickly by pressing a wedge-shaped tool into soft clay tablets, the spacing wasn't always consistent. Knowing the actual value often depended on understanding the context of what was being described. Sometime around the 4th century B.C., the Babylonians started using their end-of-sentence symbol (we'll use a dot) to show that a place was being skipped, so that 72 and 3612 became

$$\nabla \triangleleft \nabla \nabla \qquad \text{and} \qquad \nabla \cdot \triangleleft \nabla \nabla$$

respectively. Thus, zero began its life as "place holder," a symbol for something skipped.

Credit for developing the base-ten place value system we now use belongs to the people of India, sometime before 600 A.D. They used a small circle as the place-holder symbol. The Arabs learned this system in the 9th century, and their influence gradually spread it into Europe in the two or three centuries that followed. The symbols for the single digits changed a bit, but the principles remained the same. (The Arabs used the circle symbol to represent "five"; they used a dot for the place holder.) The Indian word for this absence of quantity, *sunya*, became

---

[1] See Sketch 1 for some more details about the Babylonian numeration system.

the Arabic *sifr*, then the Latin *zephirum* (along with a barely Latinized *cifra*), and these words in turn evolved into the English words *zero* and *cipher*. Today zero, usually as a circle or an oval, still indicates that a certain power of ten is not being used.

But that's only the beginning of the story. By the 9th century A.D., the Indians had made a conceptual leap that ranks as one of the most important mathematical events of all time. They had begun to recognize *sunya*, the absence of quantity, as a quantity in its own right! That is, they had begun to treat zero as a number.  For instance, the mathematician Mahāvīra wrote that a number multiplied by zero results in zero, and that zero subtracted from a number leaves the number unchanged. He also claimed that a number divided by zero remains unchanged. A couple of centuries later, Bhāskara II declared a number divided by zero to be an infinite quantity.

The main point here is not which Indian mathematician got the right answers when computing with zero, but the fact that they asked such questions in the first place. To compute with zero, you must first recognize it as *something*, an abstraction on a par with *one, two, three*, etc. That is, you must move from counting one goat or two cows or three sheep to thinking of 1, 2, and 3 as things that can be manipulated without concern for what kinds of objects are being counted. Then you must take an extra step, to think of 1, 2, 3,..., as ideas that exist *even if they aren't counting anything at all*. Then, and only then, does it make sense to treat 0 as a number. The ancient Greeks never took that extra step in abstraction; it was fundamentally at odds with their sense of a number as a quantitative property *of things*.

The Indian recognition of 0 as a number was a key for unlocking the door of algebra. Zero, as symbol and as concept, found its way to the West largely through the writings of the 9th-century Arab scholar Muḥammad ibn Mūsa al-Khwārizmī. He wrote two books, one on arithmetic and the other on solving equations, which were translated into Latin in the 12th century and circulated throughout Europe.

In al-Khwārizmī, zero is not yet thought of as a number; it is just a place holder. In fact, he describes the numeration system as using "nine symbols," meaning 1 through 9. In one of the Latin translations, the role of zero is explained like this:

> But when [ten] was put in the place of one and was made in the second place, and its form was the form of one, they needed a form for the tens because of the fact that it was similar to the form of one, so that they might know by

means of it that it was [ten]. So they put one space in front
of it and put in it a little circle like the letter o, so that by
means of this they might know that the place of the units
was empty and that no number was in it except the little
circle...[2]

The Latin translations often began with "Dixit Algorizmi," meaning "al-Khwārizmī said." Many Europeans learned about the decimal
place system and the essential role of zero from these translations. The
popularity of this book as an arithmetic text gradually led its title to
be identified with the methods in it, giving us the word "algorithm."

As the new system spread and people started to compute with the
new numbers, it became necessary to explain how to add and multiply
when one of the digits was zero. This helped make it seem more like a
number. Nevertheless, the Indian idea that one should treat zero as a
number in its own right took a long time to get established in Europe.
Even some of the most prominent mathematicians of the 16th and 17th
centuries were unwilling to accept zero as a root (solution) of equations.

However, two of those mathematicians used zero in a way that transformed the theory of equations. Early in the 17th century, Thomas Harriot, who was also a geographer and the first surveyor of the Virginia
colony, proposed a simple but powerful technique for solving algebraic
equations:

> Move all the terms of the equation to one side of the equal
> sign, so that the equation takes the form
>
> [some polynomial] = 0.

This procedure, which one author calls Harriot's Principle,[3] was popularized by Descartes in his book on analytic geometry and is sometimes
credited to him. It is such a common part of elementary algebra today that we take it for granted, but it was a truly revolutionary step
forward at the time. Here is a simple example of how it works for us:

> To find a number x for which $x^2 + 2 = 3x$ is true (a *root* of
> the equation), rewrite it as
>
> $$x^2 - 3x + 2 = 0.$$
>
> The left side can be factored into $(x - 1) \cdot (x - 2)$. Now, for
> the product of two numbers to equal 0, at least one of them
> must equal 0. (This is another special property of zero that

---

[2]From [34]. The actual text has the Roman numeral "X" for "ten."

[3]See [36], p. 301.

makes it unique among numbers.) Therefore, the roots can
be found by solving two much easier equations,

$$x - 1 = 0 \qquad \text{and} \qquad x - 2 = 0.$$

That is, the two roots of the original equation are 1 and 2.

Of course, we chose this example because it factors easily, but a lot
was known about factoring polynomials, even in Harriot's time, so this
principle was a major advance in the theory of equations.

When linked with the coordinate geometry of Descartes,[4] Harriot's Principle
becomes even more powerful. We'll use
modern terminology to explain why. To
solve any equation with real-number variable x, rewrite it as $f(x) = 0$, where $f(x)$
is some function of x. Now graph $f(x)$.
The roots (solutions) of the original equa-

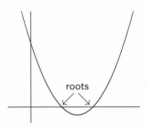

tion occur where this graph crosses the x-axis. Thus, even if the equation can't be solved exactly, a good picture of it will give you a good
approximation of its solutions.

By the 18th century, then, the status of zero had grown from place
holder to number to algebraic tool. There is one more step in this
number's claim to mathematical prominence. As 19th-century mathematicians generalized the structure of the number systems to form
the rings and fields of modern algebra, zero became the prototype for
a special element. The facts that 0 plus a number leaves that number unchanged and 0 times a number results in 0 became the defining
properties of the "additive identity" element of these abstract systems,
often called simply the *zero* of the ring or field. And the driving force
 behind Harriot's Principle — if a product of numbers is 0,
then one of them must be 0 — characterized a particularly
important type of system called an *integral domain*. Not
a bad career for a cipher, don't you think?

For a Closer Look: Many books discuss the history of zero. The
material in [36] and in [133] is particularly interesting. Also worth
noting is [95], a more literary take on the story.

---

[4]See Sketch 16 for more about coordinate geometry.

Broken Numbers

Writing Fractions

Fractions have been part of mathematics for 4000 years or so, but the way we write them and think about them is a much more recent development. In earlier times, when people needed to account for portions of objects, the objects were broken down, sometimes literally, into smaller pieces and then the pieces were counted. (Even our word "fraction," with the same root as "fracture" and "fragment," suggests breaking something up.) This evolved into primitive systems of weights and measures, which made the basic measurement units smaller as more precision was desired. In modern terms, we might say that ounces would be counted instead of pounds, inches instead of feet, cents instead of dollars, etc. Of course, those particular units were not used in early times, but their predecessors were. Some measurement systems still in use today reflect the desire to count smaller units, rather than to deal with fractional parts. For instance, in the following list of familiar liquid measures, each unit is half the size of its predecessor:

gallon, half-gallon, quart, pint, cup, gill.

(In fact, each of these units can be expressed in terms of an even smaller unit, the *fluid ounce*. A gill is 4 fluid ounces.)

Thus, in its earliest forms, the fraction concept was limited mainly to *parts*, what we today would call *unit fractions*, fractions with numerator 1. More general fractional parts were handled by combining unit fractions; what we would call *three fifths* was thought of as "the half and the tenth."[1] This limitation made writing fractions easy. Since the numerator was always 1, it was only necessary to specify the denominator and mark it somehow to show that it represented the part, rather than a whole number of things. The Egyptians did this by putting a symbol meaning "part" or "mouth" over the hieroglyphic numeral.[2] For example, using a simple dot or oval for this symbol,

ten: ∩   the tenth: ∩̇   twelve: ∩‖   the twelfth: ∩̑‖

---

[1] See pp. 30–32 of [30] for a nice explanation of this concept of "parts" in ancient Egypt and [64] or [29] for an extensive discussion.

[2] See Sketch 1 for a description of Egyptian numerals. The Egyptians also had special symbols for one half, two thirds, and three fourths.

While writing down "parts" was easy, working with them was not. Suppose we take "the fifth" and double it. We'd say we get two fifths. But in the "parts" system, "two fifths" is not legitimate; the answer must be expressed as a sum of parts; that is, the double of "the fifth" is "the third and the fifteenth." (Right?) Since the Egyptians' multiplication process depended on doubling, they produced extensive tables listing the doubles of various parts.

The Mesopotamian scribes, as usual, went their own way. They extended the sexagesimal (base-sixty) system described in Sketch 1 to handle fractions, too, just as we do with our decimal system. So, just as they would write 72 (in their symbols) as "1,12" — meaning $1 \times 60 + 12$, they would write $72\frac{1}{2}$ as "1,12;30" — meaning $1 \times 60 + 12 + 30 \times \frac{1}{60}$. That was quite a workable system. As used in ancient Babylon, however, it had one major problem: The Babylonians did not use a symbol (like the semicolon above) to indicate where the fractional part began. So, for example, "30" in a cuneiform tablet might mean "30" or it might mean "$\frac{30}{60} = \frac{1}{2}$." To decide which is really meant, one has to rely on the context.

Both the Egyptian and the Babylonian systems were passed on to the people of Greece, and from them to the Mediterranean cultures. Greek astronomers learned about sexagesimal fractions from the Babylonians and used them in their measurements — hence degrees, minutes, and seconds. This remained common in technical work. Even when the decimal system was adopted for whole numbers, people still used sexagesimals for fractions.

In everyday life, however, people in Greece used a system very similar to the Egyptian system of "parts." In fact, the practice of dealing with fractional values as sums or products of unit fractions dominated the arithmetic of fractions in Greek and Roman times and lasted well into the Middle Ages. (The exception is Diophantus, from around the 3rd century A.D., though scholars argue about exactly how Diophantus thought about fractions. See the Overview for more about his work. Diophantus is often an exception when it comes to Greek mathematics!) Fibonacci's *Liber Abbaci*, an influential 13th-century European mathematics text, made extensive use of unit fractions and described various ways of converting other fractions to sums of unit fractions.

There was another system in use from antiquity, also based on the notion of parts, but multiplicative. In that system, the process required taking a part *of* a part (of a part of...). For instance, in this system we might think of 2/15 as "two fifth parts of the third part." There were even constructions like "the third of two fifth parts and the third," which was intended to mean

$$\left(\frac{1}{3} \times \frac{2}{5}\right) + \frac{1}{3} = \frac{7}{15}.$$

As recently as the 17th century, Russian manuscripts on surveying referred to one ninety-sixth of a particular measure as "half-half-half-half-half-third,"[3] expecting the reader to think in terms of successive subdivisions:

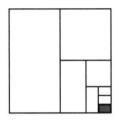

$$\tfrac{1}{3} \text{ of } \tfrac{1}{2} \text{ of } \tfrac{1}{2} \text{ of } \tfrac{1}{2} \text{ of } \tfrac{1}{2} = \tfrac{1}{96}$$

In contrast to this unit-fraction approach, our current approach to fractions is based on the idea of measuring by counting copies of a single, small enough part. Instead of measuring out a pint and a cup of milk for a recipe, it's easier to measure three cups. That is, instead of representing a fractional amount by identifying the largest single part within it and then exhausting the rest by successively smaller parts, we simply look for a small part that can be counted up enough times to produce exactly the amount we want. Two numbers would then specify the total amount: the size of the unit part, and the number of times we count it.

This was also how the Chinese mathematicians thought about fractions. The *Nine Chapters on the Mathematical Art*, which dates back to about 100 B.C., contains a notation for fractions that is very similar to ours. The one difference is that the Chinese avoided using "improper fractions" such as $\frac{7}{3}$; they would write $2\frac{1}{3}$ instead. All the usual rules for operating with fractions appear in the *Nine Chapters*: how to reduce a fraction that is not in lowest terms, how to add fractions, and how to multiply them. For instance, the rule for addition (translated into our terminology) looks like this:

> Each numerator is multiplied by the denominators of the other fractions. Add them as the dividend, multiply the denominators as the divisor. Divide; if there is a remainder, let it be the numerator and the divisor be the denominator.[4]

This is pretty much what we still do!

For multiplying and dividing, the method explained in the *Nine Chapters* also used a kind of reduction to a common "denominator."

---

[3] See [91], p. 33, [164], Vol II, p. 213.

[4] [159], p. 70.

This made the process of division natural and obvious. For example, to divide $\frac{2}{3}$ by $\frac{4}{5}$, they would first multiply both the numerator and denominator of each fraction by the denominator of the other, so that

$$\frac{2}{3} \div \frac{4}{5} \quad \text{becomes} \quad \frac{2 \cdot 5}{3 \cdot 5} \div \frac{3 \cdot 4}{3 \cdot 5}; \quad \text{that is,} \quad \frac{10}{15} \div \frac{12}{15}.$$

Now that both fractions are written in the same "unit of measurement" (denominator), the question is reduced to a whole-number division problem: dividing the numerator of the first fraction by the numerator of the second. In this case, then,

$$\frac{2}{3} \div \frac{4}{5} = \frac{2 \cdot 5}{3 \cdot 5} \div \frac{3 \cdot 4}{3 \cdot 5} = \frac{10}{15} \div \frac{12}{15} = 10 \div 12 = \frac{5}{6}.$$

A similar approach (perhaps learned from the Chinese) appears in manuscripts from India as early as the 7th century A.D. They wrote the two numbers one over the other, with the size of the part below the number of times it was to be counted. No line or mark separated one number from the other. For instance (using our modern numerals), the fifth part of the basic unit taken three times would be written as $\begin{smallmatrix} 3 \\ 5 \end{smallmatrix}$.

The Indian custom of writing fractions as one number over another became common in Europe a few centuries later. Latin writers of the Middle Ages were the first to use the terms *numerator* ("counter" — how many) and *denominator* ("namer" — of what size) as a convenient way of distinguishing the top number of a fraction from the bottom one. The horizontal bar between the top and bottom numbers was inserted by the Arabs by sometime in the 12th century. It appeared in most Latin manuscripts from then on, except for the early days of printing (the late 15th and early 16th centuries), when it probably was omitted because of typesetting problems. It gradually came back into use in the 16th and 17th centuries. Curiously, although 3/4 is easier to typeset than $\frac{3}{4}$, this "slash" notation did not appear until about 1850.

The form in which fractions were written affected the arithmetic that developed. For instance, the "invert and multiply" rule for dividing fractions was

$$\frac{2}{3} \div \frac{5}{7} = \frac{2}{3} \times \frac{7}{5} = \frac{14}{15}$$

used by the Indian mathematician Mahāvīra around 850 A.D. However, it was not part of Western (European) arithmetic until the 16th century, probably because it made no sense unless fractions, including fractions larger than 1, were routinely written as one number over another.

The term *per cent* ("for every hundred") as a name for fractions with denominator 100 began with the commercial arithmetic of the 15th and 16th centuries, when it was common to quote interest rates in hundredths. Such customs have persisted in business, reinforced in this country by a monetary system based on dollars and *cents* (hundredths of dollars). This has ensured the continued use of percents as a special branch of decimal arithmetic. The percent symbol evolved over several centuries, starting as a handwritten abbreviation for "per 100" around 1425 and gradually being transformed into "per $\frac{0}{0}$" by 1650, then simply to "$\frac{0}{0}$," and finally to the "%" sign we use today.[5]

While decimal fractions occur fairly early in Chinese mathematics and Arabic mathematics, these ideas do not seem to have migrated to the West. In Europe, the first use of decimals as a computational device for fractions occurred in the 16th century. They were made popular by Simon Stevin's 1585 book, *The Tenth*. Stevin, a Flemish mathematician and engineer, showed in his book that writing fractions as decimals allows operations on fractions to be carried out by the much simpler algorithms of whole number arithmetic. Stevin sidesteps the issue of infinite decimals. After all, he was a practical man writing a practical book. For him, 0.333 was as close to $\frac{1}{3}$ as you might want.

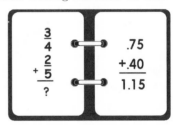

Within a generation, the use of decimal fractions by scientists such as Johannes Kepler and John Napier paved the way for general acceptance of decimal arithmetic. However, the use of a period as the *decimal point* was not uniformly adopted for many years after that. For quite a while, many different symbols — including an apostrophe, a small wedge, a left parenthesis, a comma, a raised dot, and various other notational devices — were used to separate the whole and fractional parts of a number. In 1729, the first arithmetic book printed in America used a comma for this purpose, but later books tended to favor the use of a period. Usage in Europe and other parts of the world continues to be varied. In many countries the comma is used instead of the period as the separation symbol of choice. Most English-speaking countries use the period, but most other European nations prefer the comma.

---

[5]See [22], p. 312.

International agencies and publications often accept both comma and period. Modern computer systems allow the user to choose, as one of the regionalization and language settings, whether the decimal separator should be written as a comma or a period.

Stevin's innovation, together with its application to science and practical computation, had an important effect on how people understood numbers. Up to Stevin's time, things like $\sqrt{2}$ or (even worse) π were not quite considered numbers. They were ratios that corresponded to certain geometric quantities, but when it came to thinking of them as numbers, people felt queasy. The invention of decimals allowed people to think of $\sqrt{2}$ as 1.414 and of π as 3.1416, and suddenly, to paraphrase Tony Gardiner,[6] "all numbers looked equally boring." It's no coincidence that it was Stevin who first thought of the real numbers as points on a number line and who declared that all real numbers should have equal status.

When calculators were introduced in the middle of the 20th century, it seemed as if decimals had won the day permanently. But the old system of numerators and denominators still has many advantages, both computational and theoretical, and it has proved to be remarkably resilient. We now have calculators and computer programs that can work with common fractions. Percentages are used in commerce, common fractions and mixed numbers appear in recipes, and decimals occur in scientific measurements. These multiple representations are a matter of convenience and also a reminder of the rich history behind ideas we use every day.

For a Closer Look: For more information about this topic, see section 1.17 of [74], which also contains references to further historical literature. Also worth consulting are pages 309–335 of [22] and pages 208–250 of [164], which contain a wealth of material on the historical development of rational numbers and their various notational forms. A particularly nice account of Stevin's innovation and its problems can be found in [62] and [63].

---

[6]Former president of the (British) Mathematical Association, quoted in [160].

Less Than Nothing?
Negative Numbers

Did you know that negative numbers were not generally accepted, even by mathematicians, until a few hundred years ago? It's true. Columbus discovered America more than two centuries before negatives joined the society of numbers. They didn't become first-class citizens until the middle of the 19th century, about the time of the American Civil War.

Numbers arose from counting and measuring things: 5 goats, 37 sheep, 100 coins, 15 inches, 25 square meters, etc. Fractions were just a refined form of counting, using smaller units: $\frac{5}{8}$ in. is five *eighths* of an inch, $\frac{3}{10}$ mi. is three *tenths* of a mile, and so on. And if you're counting or measuring, the smallest possible quantity must be zero, right? After all, how can any quantity be less than nothing? Thus, it is not too surprising that the idea of a negative number — a number less than zero — was a difficult concept.

"So," you may ask, "where did this strange idea come from? How did anyone think of such numbers?" The usual answer is that negative numbers first appeared on the mathematical scene when people began to solve problems such as:

"I am 7 years old and my sister is 2. When will I be exactly twice as old as my sister?"

This translates into solving the equation $7 + x = 2(2 + x)$, where x is the number of years from now that this will happen. As you can see, in this case the answer is 3 (years from now). But the same kind of question can be asked for any ages. That is, we can as easily ask for the solution of $18 + x = 2(11 + x)$. In this case, however, the solution is negative: $x = -4$. And this answer even makes sense: if I am now 18 and my sister is 11, I was twice as old as she was four years ago.

In fact, however, negative numbers did *not* first appear in that context. They appeared as *coefficients* long before they appeared as answers. For a long time negative answers were just regarded as nonsensical. Later, they were seen as a signal that the question was wrongly posed. In our example, the negative solution would have been understood as revealing that I asked the wrong question; it should have been "how long ago?" (a positive number of years), not "when?".

The scribes of Egypt and Mesopotamia could solve such equations more than three thousand years ago, but they never considered the

possibility of negative solutions. Chinese mathematicians, on the other hand, had a method of solution based on manipulating the coefficients of the equations, and they seem to have been able to handle negative numbers as intermediate steps in that process.

Our mathematics, like much of our Western culture, is rooted mainly in the work of ancient Greek scholars. Despite the depth and subtlety of their mathematics and philosophy, the Greeks ignored negative numbers completely. Most Greek mathematicians thought of "numbers" as being positive whole numbers and thought of line segments, areas, and volumes as different kinds of magnitudes (and therefore *not* numbers). Even Diophantus, who wrote a whole book about solving equations, never considered anything but positive rational numbers. For example, Problem 2 in Book V of his *Arithmetica* leads him to the equation[1] $4x + 20 = 4$. "This is absurd," he says, "because 4 is smaller than 20." To Diophantus, $4x + 20$ meant adding something to 20, and hence could never be equal to 4. On the other hand, Diophantus did know that when we expand $(x - 2)(x - 6)$, "minus times minus makes plus." In other words, he understood how to work with negative coefficients.

A prominent Indian mathematician, Brahmagupta, recognized and worked with negative quantities to some extent as early as the 7th century. He treated positive numbers as possessions and negative numbers as debts, and also stated rules for adding, subtracting, multiplying, and dividing with negative numbers. Later Indian mathematicians continued in this tradition, explaining how to use and operate with negative as well as positive numbers. Nevertheless, they regarded negative quantities with suspicion for a very long time. Five centuries later, Bhāskara II, considered this problem:

> A fifth part of a troop of monkeys, minus three, squared, has gone to a cave; one is seen [having] climbed to the branch of a tree. Tell [me], how many are they?[2]

The equation is $(\frac{x}{5} - 3)^2 + 1 = x$, and Bhāskara correctly finds the roots, 50 and 5. But then he says "In this case, the second [answer] is not to be taken due to its inapplicability. For people have no confidence [or "comprehension of"] in a manifest [quantity] becoming negative." The problem is that if $x = 5$, then $\frac{x}{5} - 3 = -2$, and Bhāskara is nervous about talk of $(-2)^2$ monkeys even though $(-2)^2$ is positive.

---

[1] This wasn't how he wrote it, of course; see Sketch 8.

[2] From Bhāskara's Bījagaṇita; translation by Kim Plofker in [96, p. 476].

Early European understanding of negative numbers was not directly influenced by this work. Indian mathematics first came to Europe through the Arabic mathematical tradition. Two books written by Muḥammad ibn Mūsa al-Khwārizmī in the 9th-century were especially influential. However, the Arab mathematicians did not use negative numbers. Perhaps this is in part because algebraic symbolism as we know it today did not exist in those times. (See Sketch 8.) Problems that we solve with algebraic equations were solved completely in words, often with geometric interpretations of all the numerical data as line segments or areas. Al-Khwārizmī, for example, recognized that a quadratic equation can have two roots, but only when both of them are positive. This may have resulted from the fact that his approach to solving such equations depended on interpreting them in terms of areas and side lengths of rectangles, a context in which negative quantities made no sense. (See Sketch 10.)

One thing that the Arabs (and also Diophantus) did understand was how to expand products of the form

$$(x - a)(x - b).$$

They knew that in this situation negative times negative is positive, and negative times positive is negative. But they expected the answers to any problem to be positive. So, while these "laws of signs" were known, they weren't understood as rules about how to operate with independent things called "negative numbers."

Thus, European mathematicians learned from their predecessors a kind of mathematics that dealt only with positive numbers. Except for the hint about multiplication of negative numbers offered by the distributive law, they were left to deal with negative quantities on their own, and they proceeded much more slowly than their counterparts in India and China.

European mathematics made tremendous leaps after the Renaissance, motivated by astronomy, navigation, physical science, warfare, commerce, and other applications. In spite of that progress, and perhaps because of its utilitarian focus, there was continued resistance to negative numbers. In the 16th century, even such prominent mathematicians as Cardano in Italy, Viète in France, and Stifel in Germany rejected negative numbers, regarding them as "fictitious" or "absurd." When negatives appeared as solutions to equations, they were called "fictitious solutions" or "false roots." But by the early 17th century, the tide was beginning to turn.

As the usefulness of negative numbers became too obvious to ignore, some European mathematicians began to use negative numbers in their work.

Nevertheless, misunderstanding and skepticism about negative quantities persisted. As methods for solving equations became more sophisticated and algorithmic in the 16th and 17th centuries, a further complication added to the confusion. If negatives are accepted as numbers, then the rules for solving equations can lead directly to square roots of negatives. For instance, the quadratic formula applied to the equation $x^2 + 2 = 2x$ yields the solutions $1 + \sqrt{-1}$ and $1 - \sqrt{-1}$. (It was even worse with the formula for cubic equations; see Sketch 11.) But, if negative numbers make sense at all, then the necessary rules of their arithmetic require that squares of negatives be positive. Since squares of positive numbers must also be positive, this means that $\sqrt{-1}$, a number whose square is -1, can be neither positive nor negative!

Faced with this apparent absurdity, it was tempting for mathematicians to regard negative numbers as suspicious characters in the world of arithmetic. Early in the 17th century, Descartes called negative solutions (roots) "false" and solutions involving square roots of negatives "imaginary." In Descartes's words, the equation $x^4 = 1$, for example, has one true root ($+1$), one false root (-1), and two imaginary roots ($\sqrt{-1}$ and $-\sqrt{-1}$). (See Sketch 17 for the story of imaginary and complex numbers.) Moreover, Descartes's use of coordinates for the plane did not use negative numbers in the way that the familiar Cartesian coordinate system (named for him) does now. His constructions and calculations were concerned primarily with positive numbers, particularly with lengths of line segments; the idea of a negative x- or y-axis simply doesn't appear. (See Sketch 16 for more about this.)

In fact, even those 17th century mathematicians who accepted negative numbers were unsure of where to put them in relation to the positive numbers. Antoine Arnauld argued that, if -1 is less than 1, then the proportion $-1 : 1 = 1 : -1$, which says that a smaller number is to a larger as the larger number is to the smaller, is absurd. John Wallis claimed that negative numbers were larger than infinity. In his *Arithmetica Infinitorum* of 1655, he argued that a ratio such as $3/0$ is infinite, so when the denominator is changed to a negative (-1, say), the result must be even larger, implying in this case that $3/-1$, which is -3, must be greater than infinity.

None of these mathematicians had any difficulty with how to operate with negative numbers. They could add, subtract, multiply, and divide with them without problems. Their difficulties were with the concept itself.

Isaac Newton, in his 1707 algebra textbook *Universal Arithmetick,* didn't help. He said that "Quantities are either Affirmative, or greater than nothing, or Negative, or less than nothing."[3] Because it carried the authority of the great Sir Isaac, this definition was taken very seriously indeed. But how could any *quantity* be less than nothing?

Despite all this, by the middle of the 18th century, negatives had become accepted as numbers, more or less, in an uneasy alliance with the familiar whole numbers and positive rationals, the much-investigated irrationals, and the highly irregular complex numbers. Even so, many reputable scholars still had misgivings about them. Around  the time of the American and French Revolutions, the famous French *Encyclopédie ou Dictionnaire Raisonné des Sciences, des Arts, et des Métiers* (from which the "Encyclopedists" get their name) said, somewhat grudgingly,

> ... the algebraic rules of operation with negatives numbers are generally admitted by everyone and acknowledged as exact, whatever idea we may have about these quantities.[4]

Leonhard Euler seemed comfortable with negative quantities. In his *Elements of Algebra*, published in 1770, he says:

> Since negative numbers may be considered as debts, because positive numbers represent real possessions, we may say that negative numbers are less than nothing. Thus, when a man has nothing of his own, and owes 50 crowns, it is certain that he has 50 crowns less than nothing; for if any one were to make him a present of 50 crowns to pay his debts, he would still be only at the point nothing, though really richer than before.[5]

On the other hand, when he has to explain why the product of two negative numbers is positive, he drops the interpretation of negative numbers as debts and argues in a formal way, saying that $-a$ times $-b$ should be the opposite of a times $-b$.

Nevertheless, half a century later there were still doubters, even in the highest ranks of the mathematical community, particularly in

---

[3]From the 1728 English translation, quoted in [143], p. 192; for the Latin original see [176], volume V, p. 58.

[4]Jean le Rond d'Alembert, article on "Negative Numbers," quoted in [107], p. 597

[5][51], pp. 4-5.

England. In 1831, at the dawning of the age of steam locomotives, the
great British logician Augustus De Morgan wrote:

> The imaginary expression $\sqrt{-a}$ and the negative
> expression $-b$ have this resemblance, that either
> of them occurring as the solution of a problem in-
> dicates some inconsistency or absurdity. As far as
> real meaning is concerned, both are equally imag-
> inary, since $0 - a$ is as inconceivable as $\sqrt{-a}$.[6]

This represents the last gasp of a fading tradition in the face of
the emergence of a much more abstract approach to algebra and the
structure of the number system. With the work of Gauss, Galois, Abel,
and others in the early 19th century, the study of algebraic equations
evolved into a study of algebraic systems — that is, systems with
arithmetic-like operations. In this more abstract setting, the "real"
meaning of numbers became less important than their operational relationships to each other. In such a setting, negative numbers — numbers that were the additive opposites of their positive counterparts —
became critically important components of the number system, and
doubts about their legitimacy simply disappeared.

Ironically, this move to abstraction paved the way for a true acceptance of the usefulness of negative numbers in a variety of real-world
settings. In fact, negative numbers are routinely taught as a fundamental part of elementary school arithmetic. We take them so much for
granted that it is sometimes difficult to understand students' struggles
with what they are and how to manipulate them. Perhaps a little sym-
pathy is in order; some of the best mathematicians in history shared
those same struggles and frustrations.

For a Closer Look: Most full-length histories of mathematics include
a discussion of the history of negative numbers. For a book-length
discussion of the debate about negative numbers in England between
the 16th and 18th centuries, see [143].

---

[6]Quoted in [107], p. 593, from De Morgan's book, *On the Study and Difficulties of Mathematics.*

By Tens and Tenths
Metric Measurement

A s soon as humans began to trade, it became important to measure — how much grain, how large a horse, how long a rope, etc. Systems of measurement had to be based on some agreed-upon unit of measure. These agreed-upon units became the standards for systems of measurement, which varied from place to place and from time to time.

Some of the earliest standards were parts of the human body, such as the *span* (the distance from the tip of the thumb to the tip of the little finger with the fingers spread out), the *palm* (the breadth of the four fingers held close together), the *digit* (the breadth of the first finger or the middle finger), and the *foot*.

The problem with "standards" like these is, of course, that the sizes of human body parts vary from person to person. Thus, it was natural to choose a king or some other prominent person on whom the standard units would be based. In 12th-century England, King Henry I declared a *yard* to be the distance from the tip of his nose

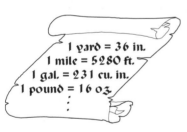

1 yard = 36 in.
1 mile = 5280 ft.
1 gal. = 231 cu. in.
1 pound = 16 oz.
:

to the tip of his thumb with his arm stretched out. This became the basis for length in the English system of measurement, a system still commonly used in the United States (and almost nowhere else). The main difficulty with using this system is calculating with the peculiar relationships among the various sizes of its units.

Many different measurement systems were used in different countries throughout the world until the late 18th century. As international trade increased, the need for a single, universally accepted standard became more and more pressing. In 1790, Bishop Charles Maurice de Talleyrand proposed to the French National Assembly a system based on the length of a pendulum that would make one full swing per second.

The French Academy of Sciences studied this plan and, after some debate, decided that the variations in temperature and gravity in different parts of the world would make this length unreliable. They proposed a new system, based on the length of a sea-level meridian arc from the Equator to the North Pole. They called one ten-millionth of this arc a *meter*.[1] (They even specified a particular meridian — the one that passes through Dunkirk, France, and Barcelona, Spain.)

 The utility of the metric system comes from the fact that all smaller and larger units of length are power-of-ten multiples of the meter. Moreover, they are named with prefixes that tell you which power of ten to use (if you know a little Latin and Greek). For the most part, units smaller than a meter have Latin prefixes; units larger than a meter have Greek ones:

$$\vdots$$

$$
\begin{aligned}
\text{gigameter} &= 1{,}000{,}000{,}000 = 10^9 \text{ meters} \\
\text{megameter} &= 1{,}000{,}000 = 10^6 \text{ meters} \\
\text{kilometer} &= 1000 \text{ meters} \\
\text{hectometer} &= 100 \text{ meters} \\
\text{dekameter} &= 10 \text{ meters} \\
\text{meter} &= 1 \text{ meter} \\
\text{decimeter} &= .1 \text{ meter} \\
\text{centimeter} &= .01 \text{ meters} \\
\text{millimeter} &= .001 \text{ meters} \\
\text{micrometer} &= .000001 = 10^{-6} \text{ meters} \\
\text{nanometer} &= .000000001 = 10^{-9} \text{ meters}
\end{aligned}
$$

$$\vdots$$

Units of area and volume come from the same basic unit, starting with the square meter and the cubic meter, respectively. Because our system of numeration is also based on powers of ten, this decimal structure makes measure conversion easy.

Perhaps less well known is the fact that units of mass (weight, more or less) in this system are also based on the meter. The *gram* was originally the basic unit, defined as the mass of one cubic centimeter of water (at the melting-point temperature of ice). More recently, the *kilogram* has become the  basic unit; it is the mass of pure water contained in a cube one decimeter on a side. This volume measure is called a *liter*. In keeping with the prefix convention, a *gram* is one thousandth of a kilogram.

---

[1]Actually, they spelled it "metre"; it comes from the Greek word *metron*, a measure.

In order for the French Academy's definition of a meter to become a practical tool, it was necessary to determine precisely the length of the meridian arc they specified. Two French surveyors were commissioned for that task. Now, up to that time, the most common unit of angular measure had been the degree. However, in the spirit of the French Academy's desire to use powers of ten, the surveyors defined a new unit of angular measure, the *grade* (or *grad*), which they declared to be one hundredth of a right angle. By this measure, coming around full circle would be 400 grades.

The surveyors used their new unit of angular mea- sure to determine the precise length of one meter. But when the French Academy accepted their findings for that length, it rejected the angular unit that helped to produce it! It chose, instead, the *radian* as the standard metric unit of angular measure, even though power-of-ten multiples of radians bear no convenient relationship to any common angles. In the judgment of the Academy, the inconvenience of departing from powers of ten was a small price to pay for preserving the unique relationship between an angular measure a in radians and the linear measure s of its arc, which is given by

$$a = \frac{s}{r}$$

where r is the radius of the circle being considered. Thus, the grade, which is sometimes called the "oldest metric unit," is not officially a metric unit at all! Nevertheless, it was used in the 19th century as a common unit of angular measure.

The Republic of France officially adopted the French Academy's system in 1795. Platinum standard models of a meter-length bar and

 a kilogram weight were constructed in 1799 and deposited in the French National Archives. However, as you might expect, public acceptance of this new system was neither immediate nor easy, even in France. It was discarded by Napoleon in 1812, but restored as the mandatory French system in 1840. International implementation became a reality when seventeen countries signed the Treaty of the Meter in 1875.

In 1960, recognizing modern technology's increasing demands for precision, the 11th General Conference on Weights and Measures established a new international system of units closely akin to the traditional metric system. This new system, called the International System of Units (SI), is based on meters and kilograms, along with five other basic units to measure time, temperature, etc., but it has redefined some of these units more precisely than before. For instance, the meter was redefined as 1,650,763.73 times the wavelength of the radiation emitted at a particular energy level by krypton-86. In 1983, the meter was redefined again, this time as the distance traveled by light in a vacuum during $\frac{1}{299,792,458}$ of a second. This may not seem like much of an improvement over measuring a meridian arc of the Earth, but it has the advantage of being reproducible in a good laboratory anywhere in the world.

The United States passed a law in 1866 making it legal (but not mandatory) to use the metric system in commerce. The U.S. was also the only English-speaking country to sign the Treaty of the Meter in 1875. However, transition from the English system to the metric system in this country has been slow and grudging. Despite the Metric Conversion Act of 1975 which urged "voluntary conversion to the metric system," there has been relatively little progress. In the last quarter of the 20th century, the ebb and flow of politics hampered wholehearted implementation of that 1975 law. A prominent instance of the cost of this ambivalence was the failure of NASA's Mars Climate Orbiter, launched in December 1998. Nine months later, this $655,000,000 mission failed just as it reached Mars because one of the ground computers sent data in English units, instead of the required metric units! Today the United States remains the last major holdout in adopting the metric system as its official standard of measurement.

For a Closer Look: For easily accessible information about this topic, look up the metric system in any good encyclopedia. See [105] for an account of the metric system in 19th century American schools. More historical information can be found in [148], which focuses especially on the role of measurement in the physical sciences. A good summary of the complete SI system is the National Institute of Standards and Technology guide to "Constants, Units, and Uncertainty," listed under Units on its website. Among other things, you'll find complete list of prefixes corresponding to powers of ten. Another good source is the website of the U.S. Metric Association.

Measuring the Circle
The Story of π

The number we call π (pronounced "pie", like the dessert or pizza) has a long and varied history. That symbol didn't originally represent a number; it's just the Greek letter that corresponds to our letter p. But the number it now labels was well known to the ancient Greeks. Long, long ago, they and others before them recognized that circles had a special, useful property: The circumference of any circle divided by its diameter always comes out to be the same number. If we agree to call that number π, then this handy fact translates into the familiar formula $C = \pi d$.

In other words, the ratio of the circumference to the diameter of a circle is always the same. We think of it as a *constant* — a number that stays the same regardless of how the other numbers in the situation may change. The scholars of ancient times also knew that this same constant ratio showed up in another basic property of circles: The area inside a circle is always that constant times the square of the radius. That is, $A = \pi r^2$. In particular, if a circle has radius 1 unit (inch, foot, meter, mile, light year, or whatever), then the area inside that circle is exactly equal to π units.

Because the circle shape is so important for so many things we humans make and use, from wheels and gears to clocks and rockets and telescopes, the constant in these two formulas is a number well worth knowing. But what is it, exactly?

From a historical viewpoint, the fascinating, troubling word here is "exactly." Finding its value has been a mystery that people of many different civilizations have worked on and puzzled over for hundreds of years. Here are a few examples:

ca. 1650 B.C.— The Rhind Papyrus, from ancient Egypt, approximated the area of a unit circle as $4(\frac{8}{9})^2$.

ca. 240 B.C.— Archimedes showed that it is between $3\frac{10}{71}$ and $3\frac{10}{70} = 3\frac{1}{7}$. Heron later popularized the use of $3\frac{1}{7}$ in many practical contexts.

ca. 150 A.D.— Ptolemy, a Greek astronomer, used $\frac{377}{120}$ for it.

ca. 480 — The Chinese scholar Zu Chongzhi used $\frac{355}{113}$ for it.

ca. 500 — The Indian mathematician Āryabhaṭa used $\frac{62832}{20000}$ for it.

ca. 1600 — A decimal value for it was computed to 35 places.

1706 — William Jones, a British mathematician, first used the Greek letter π as the name of this number. That symbol was adopted by the great Swiss mathematician Leonhard Euler in his publications of the 1730s and 1740s, and by the end of the century it had became the common name of this constant.

1873 — William Shanks of England computed, by hand, a decimal value for π to 607 places. It took him more than 15 years. Digits after the 527th are incorrect, but no one noticed the error for almost a century!

1949 — John von Neumann used the U.S. government's ENIAC computer to work out π to 2035 decimal places (in 70 hours).

1987 — Prof. Yasumasa Kanada of the University of Tokyo worked out π to 134,217,000 decimal places on an NEC SX-2 supercomputer.

1991 — Gregory and David Chudnovsky calculated π to 2,260,321,336 decimal places in 250 hours, using a home-built supercomputer in their New York City apartment. (This many digits, printed in a single line of ordinary newspaper type, would stretch from New York to Hollywood, California!)[1]

2002 — Prof. Kanada's team computed π to 1,241,100,000,000 decimal places! This number is almost 550 times as long as the one the Chudnovskys found. As a line of ordinary type, it would extend more than 1,500,000 miles — more than three trips to the Moon and back.

Yet *none* of these results is the *exact* value of π.

About 1765 (when America was working up to the Revolutionary War), a German mathematician named Johann Lambert proved that

---

[1]The story of the Chudnovskys and their amazing machine appears in "The Mountains of Pi," a profile by R. Preston published in *The New Yorker* in 1992. See [142].

π is an *irrational number*; that is, it cannot be expressed exactly as a common fraction (the ratio of two whole numbers). Among other things, this means that no decimal expression, no matter how far it is extended, will ever *exactly* equal π. But we can find decimals as close as we want, if we're willing to be patient and do enough work.

In fact, just a few decimal places are good enough for almost all practical purposes. The same is true for many of the approximations that were used even before decimals were invented. To illustrate this, we'll calculate the circumference of a circular lake with a diameter of exactly 1 kilometer (about 0.62 miles) using the historical approximations of π listed above, and compare the results to what we get using a modern calculator:

Source	Circumference	Approx. Difference
modern calculator	3.141592654 km	
Ahmes (1650 B.C.)	3.160493827 km	18.9 m. ($\approx$ 20 yd.)
Archimedes (240 B.C.)	3.141851107 km	28.8543 cm ($<$ 1 ft.)
Ptolemy (150)	3.141666667 km	7.4103 cm ($\approx$ 3 in.)
Zu Chongzhi (480)	3.14159292 km	.266 mm ($\approx \frac{1}{100}$ in.)
Āryabhaṭa (500)	3.1416 km	7.346 mm ($\approx \frac{1}{6}$ in.)

Even the crudest of these approximations, from more than 3600 years ago, is off by less than 2%. The rest of them miss the "real" circumference of the lake by a truly insignificant amount. So why do people bother to calculate π to thousands or millions or billions of decimal places? Is there any possible value to spending all that time and effort? Maybe so. There are many deep questions about irrational numbers that we cannot yet answer. We can prove that their decimal expansions are infinite and do not repeat any finite sequence of digits without interruption from some point on. But is there some subtle pattern in this sequence of digits? Do all ten digits appear with equal frequency, or do certain digits occur more often than others? Do certain strings of digits appear in some predictable way?

We do not even know enough yet to know precisely what questions are worthwhile. Sometimes a seemingly insignificant point leads the way to broad, new insights. And then there are questions about the hardware and software being used to generate these immense strings of digits: How can we make their capacities bigger, their speed faster, their accuracy more reliable? A problem such as generating the digits of π provides a proving ground for technological improvement.

Nevertheless, probably the most honest explanation of such persistence is simple human curiosity about the unknown. Virtually any problem without an easy solution will lure at least a few people to pursue it, sometimes obsessively. The history of both the progress and the folly of the human race is dotted with the achievements and the misadventures of such people. Not knowing in advance which questions will lead which way adds a risk factor that makes them more inviting. In mathematics, as in any sport, overcoming the challenges of the untried and the unknown is its own reward.

π = 3.1415926535897932384626433832795028841971693993 75105820974944592307816406286208998628034825342117067 98214808651328230664709384460955058223172535940812848 11174502841027019385211055596446229489549303819644288 10975665933446128475648233786783165271201909145648566 92346034861045432664821339360726024914127372458700660 63155881748815209209628292540917153643678925903600113 30530548820466521384146951941511609433057270365759595 19530921861173819326117931051185480744623799627495673 51885752724891227938183011949129833673362440656664308 60213949463952247371907021798609437027705392171762931 76752384674818467669405132000568127145263560827785771 34275778960917363717872146844090122495343014654958537 10507922796892589235420199561121290219608640344181598 13629774771309960518707211349999998372978049951059731 73281609631859502445945534690830264252230825333468503 52619311881710100031378387528865875332083814206171776 69147303598253490428755468731159562863882353787593751 95778185778053217122680661300192787661119590921642019 89 . . .

The first 1000 decimal places of π

For a Closer Look: Beckmann's [11] is a readable book about the history of π. Also worth looking at is [14], which collects many articles, including some of the original sources (for example, it contains a sampling from Shanks's original publication). The latest information about Prof. Kanada's computations can be found at the Kanada Laboratory home page.

8 The Cossic Art
Writing Algebra with Symbols

When you think of *algebra*, what comes to mind first? Do you think of equations or formulas made up of x's and y's and other letters, strung together with numbers and arithmetic symbols? Many people do. In fact, many people regard algebra simply as a collection of rules for manipulating symbols that have something to do with numbers.

There's some truth in that. But describing algebra solely in terms of its symbols is like describing a car by its paint job and body style. What you see is *not* all you get. In fact, like a car, most of what makes algebra run is "under the hood" of its symbolic appearance. Nevertheless, just as an automobile's body styling can affect its performance and value, so does the symbolic representation of algebra affect its power and usefulness.

An algebra problem, regardless of how it's written, is a question about numerical operations and relations in which an unknown quantity must be deduced from known ones. Here's a simple example:

> Twice the square of a thing is equal to five more than three times the thing. What is the thing?

Despite the absence of symbols, this is clearly an algebra question. Moreover, the word "thing" was a respectable algebraic term for a very long time. In the 9th century, al-Khwārizmī (whose book title, *al-jabr w'al muqābala*, is the source of the word "algebra") used the word *shai* to mean an unknown quantity. When his books were translated into Latin, this word became *res*, which means "thing". For instance, John of Seville's 12th-century elaboration of al-Khwārizmī's arithmetic contains this question, which begins "Quaeritur ergo, quae res...":[1]

> It is asked, therefore, what thing together with 10 of its roots or what is the same, ten times the root obtained from it, yields 39.

In modern notation, this would be written either as $x + 10\sqrt{x} = 39$ or as $x^2 + 10x = 39$. (An "X" appears in the Latin version of this question, but it's actually the Roman numeral for 10. To avoid such confusions and emphasize more significant variations in notation, we use familiar numerals in all these algebra examples.)

---

[1]See p. 336 of [22] for both the original Latin and this translation.

Some Latin texts used *causa* for Al-Khwārizmī's *shai*, and, when these books were translated into Italian, *causa* became *cosa*. As other mathematicians studied these Latin and Italian texts, the word for the unknown became *Coss* in German. The English picked up on this and called the study of questions involving unknown numbers "the Cossic Art" (or "Cossike Arte" in the spelling of those days) — literally, "the Art of Things".

Like most of our familiar algebraic symbols, the x and other letters we now use to represent unknown numbers are relative newcomers to the "art." Many early symbols were just abbreviations for frequently used words: p or $\tilde{p}$ or $\bar{p}$ for "plus," m or $\tilde{m}$ or $\bar{m}$ for "minus," and so on. They saved writing time and print space, but they did little to promote a deeper understanding of the ideas they expressed. Without consistent and illuminating symbolism, algebra was indeed an art, an often idiosyncratic activity heavily dependent on the skill of its individual practitioners. Just as standardization of parts was a critical step in the mass production of Henry Ford's automobiles, so the standardization of notation was a critical step in the use and progress of algebra.

Good mathematical notation is far more than efficient shorthand. Ideally, it should be a universal language that clarifies ideas, reveals patterns, and suggests generalizations. If we invent a really good notation, it sometimes seems to think for us: just manipulating the notation achieves results. As Howard Eves once said, "A formal manipulator in mathematics often experiences the discomforting feeling that his pencil surpasses him in intelligence."[2]

Our current algebraic notation is close to this ideal, but its development has been long, slow, and sometimes convoluted. For a flavor of that development, we'll look at various ways in which a typical algebraic equation would likely have been written in different times and places during the progress of algebra in Europe. (To highlight the notational development, we use English in place of Latin or other languages when words, rather than symbols, would be used.)

Here is an equation containing some common ingredients of early algebraic investigations:

$$x^3 - 5x^2 + 7x = \sqrt{x + 6}$$

In 1202, Leonardo of Pisa would have written that equation (perhaps rearranged for clarity) entirely in words, something like this:

> The cube and seven things less five squares is equal to the root of six more than the thing.

---

[2]See [55], entry 251.

This approach to writing mathematics is usually called *rhetorical*, in contrast to the symbolic style we use today. In the 13th and 14th centuries, European mathematics was almost entirely rhetorical, with occasional abbreviations here and there. For instance, Leonardo began to use R for "square root" in some of his later writings.

Late in the 15th century, some mathematicians started to use symbolic expressions in their work. Luca Pacioli, whose *Summa de Aritmetica* of 1494 served as a main source of Europe's introduction to the cossic art, would have written

$$cu.\tilde{m}.5.ce.\tilde{p}.7.co.\text{———}\mathcal{R}v.co.\tilde{p}.6.$$

In this notation, *co* is an abbreviation for "cosa," the unknown quantity. The abbreviations *ce* and *cu* are for "censo" and "cubo," words that the Italian mathematicians used for the square and the cube of the unknown, respectively. Notice that we refer to *the* unknown here. A fundamental weakness of this notation was its inability to represent more than one unknown in an expression. (By way of contrast, Indian mathematicians had been using the names of colors to represent multiple unknowns as early as the 7th century.) Some other interesting features of Pacioli's notation are the dots that separate each item from the next, a long dash for equality, and the symbol $\mathcal{R}$ to denote square root. The grouping of terms after the root sign was signaled by v, an abbreviation for "universale." The notation used in Girolamo Cardano's *Ars Magna* half a century later in Italy was almost identical to this.

In early 16th-century Germany, some of the symbols we use now began to appear. The $+$ and $-$ signs were adopted from commercial arithmetic and the "surd" symbol, $\sqrt{\ }$, for square root evolved, some say from a dot with a "tail," others say from a handwritten r. Equality was noted by abbreviating either the Latin or German word for it, and the grouping of terms (such as the sum after the $\sqrt{\ }$ sign) was signaled by dots. Thus, in Christoff Rudolff's *Coss* of 1525 (which has an impossibly long formal title) or Michael Stifel's *Arithmetica Integra* of 1544, our equation might have appeared as

$$c^e - 5\,\zeta + 7\,\mathcal{R}\ aequ.\ \sqrt{.}\mathcal{R} + 6.$$

As in the earlier Italian notation described above, different powers of the unknown had distinct, unrelated symbols. Its first power was called the root (*radix*) and represented by $\mathcal{R}$. The symbol for its square was ζ, a small script z which was the first letter of its German name, *zensus*. The third power, *cubus*, was symbolized by c^e. Higher powers of the unknown were written by combining the square and cube symbols multiplicatively, when possible; the fourth power was ⅄⅄, the sixth power was ⅄c^e, and so on. Higher prime powers were handled by introducing new symbols.

Easier ways to denote powers of the unknown had already begun to emerge in other countries. One of the most creative instances of this appeared in a 1484 manuscript by Nicholas Chuquet, a French physician. Like others of his time, Chuquet confined his attention to powers of a single unknown. However, he denoted the successive powers of the unknown by putting numerical superscripts on the coefficients. For example, to denote $5x^4$ he would write 5^4. He did a similar thing for roots, writing $\sqrt[3]{5}$ as $\mathcal{R}^3.5$. Chuquet was also well ahead of his time in treating zero as a number (particularly as an exponent) and in using an underline for aggregation. If our example equation had appeared in his manuscript, it would have looked like this:

$$1^3.\bar{m}.5^2.\bar{p}.7^1.\ montent\ \mathcal{R}^2.\underline{1^1.\bar{p}.6^0}.$$

Unfortunately for the development of algebraic notation, Chuquet's work was not published at the time it was written, so his innovative ideas were known only to a few mathematicians by the beginning of the 16th century. This system of denoting powers of the unknown reappeared in 1572 in the work of Rafael Bombelli, who placed the exponents in small cups above the coefficients. Bombelli's work was more widely known than that of Chuquet, but his notation was not immediately adopted by his contemporaries. In the 1580s it was picked up by Simon Stevin of Belgium, a military engineer and inventor, who used circles around the exponents. Stevin's mathematical writing emphasized the convenience of decimal arithmetic. Some of his publications were translated into English early in the 17th century, thereby carrying both his ideas and his notation across the English Channel.

A major breakthrough in notational flexibility and generality was made by François Viète in the last decade of the 16th century. Viète was a lawyer, a mathematician, and an advisor to King Henri IV of France with duties that included deciphering messages written in secret codes. His mathematical writings focused on methods of solving algebraic equations, and to clarify and generalize his work he introduced a revolutionary notational device. In Viète's own words:

> In order that this work may be assisted by some art, let the
> given magnitudes be distinguished from the undetermined
> unknowns by a constant, everlasting and very clear symbol,
> as, for instance, by designating the unknown magnitude
> by means of the letter A or some other vowel... and the

given magnitudes by means of the letters B, G, D or other consonants.[3]

Using letters for both constants and unknowns allowed Viète to write general forms of equations, instead of relying on specific examples in which the particular numbers chosen might improperly affect the solution process. Some earlier writers had experimented with using letters, but Viète was the first to use them as an integral part of algebra. It may well be that the emergence of this powerful notational device was delayed because the Hindu-Arabic numerals were not commonly used until well into the 16th century. Prior to that, Greek and Roman numerals were used for writing numbers, and these systems used letters of the alphabet for specific quantities.

As soon as equations contained more than one unknown, it became clear that the old exponential notation was insufficient. It would not do to write $5^3 + 7^2$ if one meant $5A^3 + 7E^2$. In the 17th century, several competing notational devices for this appeared almost simultaneously. In the 1620s, Thomas Harriot in England would have written it as $5aaa + 7ee$. In 1634, Pierre Hérigone of France wrote unknowns with coefficients before and exponents after, as in $5a3 + 7e2$. In 1636, James Hume (a Scotsman living in Paris) published an edition of Viète's algebra with exponents elevated and in small Roman numerals, as in $5a^{iii} + 7e^{ii}$. In 1637, a similar notation appeared in René Descartes's *La Géométrie*, but with the exponents written as small Hindu-Arabic numerals, as in $5a^3 + 7e^2$. Of these notations, Harriot's and Hérigone's were the easiest to typeset, but conceptual clarity won out over typographical convenience and Descartes's method eventually became the standard notation used today.

Descartes's influential work is also the source of some other notational devices that have become standard. He used lowercase letters from the end of the alphabet for unknowns and lowercase letters from the beginning of the alphabet for constants. He also used an overline bar from the $\sqrt{}$ sign to indicate its scope. However, he introduced the symbol ∞ for equality. Thus, Descartes's version of our sample equation would be very much, but not entirely, like our own:

$$x^3 \text{ -- } 5xx + 7x \propto \sqrt{x+6}$$

The $=$ sign for equality, proposed in 1557 by Robert Recorde[4] and widely used in England, was not yet popular in continental Europe. In

---

[3]From Viète's *In artem analyticam Isagoge* of 1591, as translated by J. Winfree Smith. See [106], p 340.

[4]See Sketch 2 for more details about this.

the 17th century it was only one of several different ways of symbolizing equality, including $\sim$ and the ∞ sign of Descartes. Moreover, $=$ was being used to denote other ideas at this time, including parallelism, difference, and "plus or minus." Its eventual universal acceptance as the symbol for "equals" is probably due in large part to its adoption by both Isaac Newton and Gottfried Leibniz. Their systems of the calculus dominated the mathematics of the late 17th and early 18th centuries, so their notational choices became widely known. During the 18th century, the superior calculus notation of Leibniz gradually superseded that of Newton. Had Leibniz chosen to use Descartes's symbol instead of Recorde's, we might be using ∞ for equality today.

This sketch has tried to capture the flavor of the long, erratic, sometimes perverse way in which algebraic symbolism has developed. In hindsight, "good" notational choices have proved to be powerful stimuli for mathematical progress. Nevertheless, those choices often were made with little awareness of their importance at the time. The evolution of exponential notation is a prime example of this. Powers of an unknown quantity were trapped for centuries by the limited geometric intuition of squares and cubes, and the notation reinforced this confinement. Descartes finally liberated them by treating squares, cubes, and the like as magnitudes independent of geometric dimension, giving a new legitimacy to x^4, x^5, x^6, and so on. From there the notation itself suggested natural extensions — to negative integral exponents (reciprocals), to rational exponents (roots of powers), to irrational exponents (limits of roots of powers), and even to complex exponents. And in the 20th century, this exponential notation was reconnected with the geometric idea of dimension to help lay the foundation of a new field of mathematical investigation: fractal geometry.

For a Closer Look: Joseph Mazur's [122] is a readable account of the evolution of mathematical notation. The standard surveys also discuss the topic. For information on the history of specific mathematical notations, the best reference is still [22], though *Earliest Uses of Various Mathematical Symbols*, a website maintained by Jeff Miller, is now a serious contender. For more on the history of algebra, see [100].

9

Linear Thinking
Solving First Degree Equations

Proberms that reduce to solving an equation of degree one arise naturally whenever we apply mathematics to the real world. It's not surprising, then, to find that almost everyone who studied mathematics, from the Egyptian scribes to the Chinese civil servants, developed techniques for solving such problems.

The Rhind Papyrus, a collection of problems probably used for training young scribes in Ancient Egypt, contains several problems of this kind. Some are simple and straightforward, others quite complicated. Here's a simple one:

A quantity; its half and its third are added to it. It becomes 10.

In our notation, that is just the equation

$$x + \frac{1}{2}x + \frac{1}{3}x = 10.$$

(Keep in mind, though, that this kind of symbolism was still far in the future, as explained in Sketch 8.) The scribe is instructed to solve it just as we would: divide 10 by $1 + \frac{1}{2} + \frac{1}{3}$.

Quite often, however, the problems in the Rhind Papyrus are solved by a very different method.

A quantity; its fourth is added to it. It becomes 15.

Instead of dividing 15 by $1\frac{1}{4}$, the scribe proceeds as follows. He assumes (or *posits*) that the quantity is 4. (Why 4? Because it's easy to compute a fourth of 4.) If you take 4 and add its fourth to it, you get $4 + 1 = 5$. So we wanted 15, but we got 5; we need to multiply what we got (that is, 5) by 3 to get what we wanted to get (that is, 15). So we take our guess and multiply it by 3. Our guess was 4, so the answer is $3 \times 4 = 12$.

This method is known as *false position*: we posit an answer that we don't really expect to be the right one, but which makes the computations easy. Then we use the incorrect result to find the number by which we need to multiply our guess to get the correct answer.

Symbols make this easy to understand. The equation we're solving looks like $Ax = B$. If we multiply x by a factor, so that it becomes kx, we see that

$$A(kx) = k(Ax) = kB.$$

So scaling the input by some factor scales the output by the same factor. This is what allows the method of false position to work; we

103

use our guess to find the right factor.

Throughout antiquity, the method of false position was used to solve linear equations, including some pretty complicated ones. These range all the way from practical problems to more fanciful problems with a recreational flavor.

However, this method can only be applied to equations of the form $Ax = B$. If, instead, the equation were $Ax + C = B$, then it is no longer true that multiplying x by a factor causes B to change by the same factor, and this simple version of the method breaks down. We might try subtracting C from both sides, but that isn't always as easy as it sounds, because the expression on the left side might initially be very complicated, and finding the correct constant to subtract would require us to simplify it to the form $Ax + C$.

Instead, a way was found to extend the basic idea to equations of that type without any such algebraic manipulation. It is called the method of *double false position*. It seems to have been brought to Europe byLeonardo of Pisa, who called it the *elchataym rule* in his *Liber Abbaci*. Leonardo learned it in North Africa, where it was in common use. In Arabic, its name was *hisab al-khata'yn*, which means "computing from two falsehoods." It seems to have been created by Arabic-speaking mathematicians; the oldest surviving description of the method comes from Qusta ibn Luqa, who lived in what is now Lebanon in the 9th century.

Double false position is so effective for solving linear equations that it continued to be used long after the invention of algebraic notations. In fact, since it doesn't require any algebra, it was taught in arithmetic textbooks as recently as the 19th century. Here's an example,[1] from *Daboll's Schoolmaster's Assistant*, published in the early 1800s.

A purse of 100 dollars is to be divided among four men A, B, C, and D, so that B may have four dollars more than A, and C eight dollars more than B, and D twice as many as C; what is each one's share of the money?

A modern approach to this would be to let x be the amount given to A. Then B gets $x + 4$, C gets $(x + 4) + 8 = x + 12$, and D gets $2(x + 12)$. Since the total is \$100, we get the equation

$$x + (x + 4) + (x + 12) + 2(x + 12) = 100,$$

which we then solve in the usual way.

---

[1] Taken from [19], pp. 34–35.

Instead, here's what *Daboll's* recommends: Make a first guess, say that A gets 6 dollars. Then B gets 10, C gets 18 and D gets 36. (Notice that we don't need to work out how D's amount is related to A's to do this; we just go step by step.) Adding up the amounts gives \$70 as the total; we're off by \$30.

So we try again. This time we guess a little higher, say that A gets 8 dollars. Then B gets 12, C gets 20, and D gets 40, for a total of \$80. That's still wrong, off by \$20.

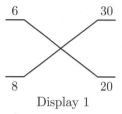

Now comes the magic. Lay out the two guesses and the two errors as in Display 1. Cross multiply: 6×20 is 120, and 8×30 is 240. Take the difference, $240 - 120 = 120$, and divide by difference of the errors, in this case by 10. The right choice for the amount A gets is $120/10 = 12$.

Display 1

This, *Daboll's* explains, is the procedure when the two errors are of the same type (both underestimates, in our case). If they were of different types, we would use the sum of the products and divide by the sum of the errors. (This is just a way of avoiding negative numbers.)

Modern readers usually find this method puzzling: Why does it work? Probably the best way to analyze it is to use some graphical thinking. No matter what the outcome of simplifying the left side of

$$x + (x + 4) + (x + 12) + 2(x + 12) = 100,$$

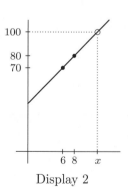

Display 2

the equation will be something of the form $mx + b = 100$. So we can think of it like this: there is a line $y = mx + b$, and we would like to determine the value of x for which $y = 100$. To determine the line, we need two points, and the two guesses provide that for us: Both $(6, 70)$ and $(8, 80)$ are on the line. We want to find x so that $(x, 100)$ is on the same line. (See Display 2.) The slope of the line is a constant; we can compute it as "rise over run" using the first and third points. We can also compute it using the second and third points, and the answers must be the same. Therefore, we see that

$$\frac{100 - 70}{x - 6} = \frac{100 - 80}{x - 8}, \quad \text{or} \quad \frac{30}{x - 6} = \frac{20}{x - 8}.$$

Notice that the numerators are exactly the errors we had before. Now cross-multiply to get

$$30(x - 8) = 20(x - 6),$$

which quickly simplifies to

$$(30 - 20)x = (30 \times 8) - (20 \times 6);$$

that is,

$$x = \frac{(30 \times 8) - (20 \times 6)}{30 - 20} = \frac{120}{10} = 12.$$

This is exactly the same computation as in the method of double false position.

Of course, our way of understanding equations as lines is quite recent (it goes back only to the 17th century; see Sketch 16), and double false position is very old. But the actual slope of the line never needs to be computed. All we need to know is that two triangles are similar, so that we have a proportion involving the lengths of the sides. That is exactly how Qusta ibn Luqa explained the rule. The crucial insight is that the change in the output is proportional to the change in the input, which is the essence of what "linearity" is all about.

The distinction between "linear" and "nonlinear" problems is still useful today. We apply it not only to equations but also to many other kinds of problems. In linear problems, there is a simple relation — a constant ratio — between changes in the input and changes in the output, exactly as we saw above. In nonlinear problems, there is no such simple relation, and sometimes very small changes in the input may produce huge changes in the output. We still don't have a complete understanding of nonlinear problems. In fact, we often use linear problems to find approximate solutions to nonlinear ones. And the methods we use for solving those linear problems are based on the same fundamental insight that serves as the basis for the method of false position.

For a Closer Look: Because solving linear equations is relatively easy, few of the standard history books have sections specifically on that subject. There is a short discussion in [19] (pp. 31–35). For more on double false position, see [156] and [155].

10 A Square and Things
Quadratic Equations

The word "algebra" comes from a title of a book written in Arabic around the year 825. The author, Muḥammad ibn Mūsa al-Khwārizmī, was probably born in what is now Uzbekistan. He lived, however, in Baghdad, which was then the world's most vibrant cultural center. Al-Khwārizmī was an able synthesizer. Starting from the practical tradition and from Indian scholarship, he wrote books on geography, astronomy, and mathematics. But his book on algebra is one of his most famous.

Al-Khwārizmī's book starts out with a discussion of quadratic equations. In fact, he considers a specific problem:

> One square, and ten roots of the same, are equal to thirty-nine dirhems. That is to say, what must be the square which, when increased by ten of its own roots, amounts to thirty-nine?

If we call the unknown x, we might call the "square" x^2. Now, a "root of this square" is x, so "ten roots of the square" is $10x$. Using this notation, the problem translates into solving the equation

$$x^2 + 10x = 39.$$

But algebraic symbolism had not been invented yet, so all al-Khwārizmī could do was to say it in words. In the time-honored tradition of algebra teachers everywhere, he follows the problem with a kind of recipe for its solution, again spelled out in words:

> The solution is this: you halve the number of the roots, which in the present instance yields five. This you multiply by itself; the product is twenty-five. Add this to thirty-nine; the sum is sixty-four. Now take the root of this, which is eight, and subtract from it half the number of the roots, which is five; the remainder is three. This is the root of the square which you sought for; the square itself is nine.[1]

Here's the computation in our symbols:

$$x = \sqrt{5^2 + 39} - 5 = \sqrt{25 + 39} - 5 = \sqrt{64} - 5 = 8 - 5 = 3.$$

---

[1] Translated by Frederic Rosen; see [2], p. 8.

It's not hard to see that this is basically just the quadratic formula as we now know it. To solve $x^2 + bx = c$, al-Khwārizmī uses the rule

$$x = \sqrt{\left(\frac{b}{2}\right)^2 + c} - \frac{b}{2}.$$

The biggest difference between this and the modern formula is that we would consider both the positive and the negative square roots. But taking the negative square root would give a negative value for x. Mathematicians at the time didn't yet believe in negative numbers; the positive root was the only one they cared about. We also put the "$-b$" part at the beginning. But that would again mean a negative number, so he prefers to put it at the end, as a subtraction. (See Sketch 5.) Finally, he states the equation with the c on the right-hand side, while we would write it as $x^2 + bx - c = 0$.

If we put the "$-b$" in front, add $\pm$ to the root, remember to take into account the sign of his c, and do a little algebra, his formula becomes ours:

$$x = -\frac{b}{2} \pm \sqrt{\left(\frac{b}{2}\right)^2 + c} = \frac{-b \pm \sqrt{b^2 + 4c}}{2}.$$

(The coefficient a is missing from this formula because al-Khwārizmī was considering a single square; that is, $a = 1$.)

But he didn't leave it at that. He felt he should explain why his method worked. Rather than doing this algebraically, as we might today, he did it with a geometric argument. It went like this:

First, we have "a square and ten roots." To picture this, draw a square whose side we don't yet know. If we call the side x, the area of the square is x^2. To get $10x$, we draw a rectangle with one side equal to x and one side equal to 10, as in Display 1.

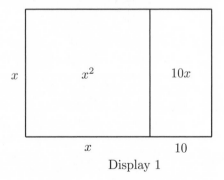

Display 1

The equation tells us that the area of the whole figure is 39. To solve the equation, that is, to determine x, we first cut the number of roots in half. Geometrically, that means we split the rectangle into two halves, each with area $5x$, as in Display 2.

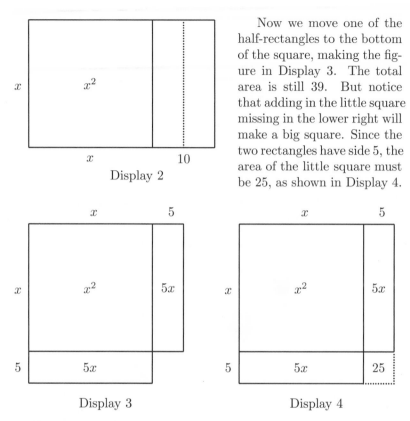

Now we move one of the half-rectangles to the bottom of the square, making the figure in Display 3. The total area is still 39. But notice that adding in the little square missing in the lower right will make a big square. Since the two rectangles have side 5, the area of the little square must be 25, as shown in Display 4.

Display 2

Display 3   Display 4

When we *complete the square* by adding in the little square, our figure becomes a square whose area is $39 + 25 = 64$. But this means that its side is equal to the square root of 64, which is 8. And since the side of the big square is $x + 5$, we can conclude that $x + 5 = 8$. So we subtract 5 and get $x = 3$.

Each step in al-Khwārizmī's rule corresponds to a step in the geometric version. And the geometric version shows us exactly what is going on and why it works! As noted before, this version of the quadratic formula assumes that the leading coefficient is 1. Today, we write the general quadratic equation as $ax^2 + bx + c = 0$, allowing for a different leading coefficient. Al-Khwārizmī would have dealt with this simply by dividing through by a and applying his rule.

After al-Khwārizmī's time many other mathematicians wrote about quadratic equations. Their methods and their geometric justifications became more and more sophisticated. But the basic idea never changed. In fact, even the example stayed the same. From the 9th century to the 16th century, books on algebra almost always started their discussion of quadratic equations by considering "a square and ten roots are equal to 39."

Early in the 17th century, mathematicians came up with the idea of using letters to represent numbers. (See Sketch 8.) Descartes suggested the convention we still use: letters from the end of the alphabet would denote *unknown* numbers, and letters from the beginning of the alphabet would stand for *known* numbers. Also, by that time resistance to the idea of negative numbers was beginning to fade. (See Sketch 5.) Thomas Harriot and René Descartes noticed that it's much easier to write all equations as something $= 0$, which we can do as long as we allow some of the coefficients to be negative. The main advantage is that $ax^2 + bx = c$, $ax^2 + c = bx$, and $ax^2 = bx + c$ could then be seen as special cases of the general equation

$$ax^2 + bx + c = 0.$$

This reduced al-Khwārizmī's three different cases to just one. Moreover, while negative solutions were still regarded with suspicion, it was at least possible to consider them. This meant that there were two square roots to account for, so the general solution could be written as

$$x = \frac{-b \pm \sqrt{b^2 - 4ac}}{2a}.$$

That's what we still do today.

For a Closer Look: There is a lot of information on the history of algebra in both [99] and [76]. Al-Khwārizmī's text can be found in many sourcebooks, including [59]. For the complete text, see [2], which can easily be found online. On the history of algebra, see [158] for a short account of the story up to the Renaissance and [100] for the full picture.

11 Intrigue in Renaissance Italy
Solving Cubic Equations

Mathematical problems rarely arise in abstract form. The problem of solving cubic equations (equations of degree 3) grew out of geometric problems first considered by Greek mathematicians. The original problems may go back as far as 400 B.C., but the complete solution only came some 2000 years later.

The story begins with a famous geometric question: Given an angle, is there a way to construct an angle one third as large? To make sense out of this question, we first need to understand (or decide) what "construct" means. If it means using only a ruler and a compass, the answer is that it cannot be done. If we allow other tools, it can. Several constructions were known in Ancient Greece, many of them involving conic sections such as parabolas and hyperbolas.

Once trigonometry was developed, it became clear that this problem boils down to solving a cubic equation, as follows. To find one third of a given angle θ, we can begin by thinking of θ as three times the angle we're looking for, which we'll call α; that is, $\alpha = \theta/3$. Now we apply the formula for the cosine of 3α:

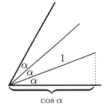

$$\cos(3\alpha) = 4\cos^3(\alpha) - 3\cos(\alpha).$$

Since the angle θ is known, we also know $\cos(\theta)$; call it a. To construct $\theta/3$, we need to construct its cosine. If we let $x = \cos(\theta/3)$, then, using the formula above with $\alpha = \theta/3$, we get $a = 4x^3 - 3x$, or $4x^3 - 3x - a = 0$. Finding x amounts to solving this equation.

When the Arabic mathematicians had begun doing algebra, it was inevitable that someone would try to apply the new techniques to equations of degree 3. The most famous mathematician to attempt this was 'Umar al-Khāyammī, known in the West as Omar Khayyám. Al-Khāyammī, who was born in Iran in 1048 and died in 1131, was famous in his time as a mathematician, scientist, and philosopher. He seems also to have been a poet, and that is how he is best known today.[1]

Because the Arabic mathematicians did not use negative numbers and did not allow zero as a coefficient, al-Khāyammī had to consider

---

[1] His most famous poetic work is the Rubā'iyāt, meaning "quatrains," which was (very freely) translated in 1859 by Edward FitzGerald as *The Rubáiyát of Omar Khayyám.*

many cases. For him, $x^3 + ax = b$ and $x^3 = ax + b$ were different kinds of equations. Arabic algebra was expressed entirely in words, so he described them as "a cube and roots are equal to a number" and "a cube is equal to roots and a number," respectively. Considered in this way, there are fourteen different kinds of cubic equations. For each of them, al-Khāyammī found a geometric solution: a construction that yields a line segment whose length satisfies the equation. Most of these constructions involve intersecting conic sections, and many have side conditions to guarantee the existence of positive solutions.

Al-Khāyammī's work is impressive, but when it comes to determining a *number* that solves the equation it is of no help at all, as he himself acknowledges. That problem was left for others to attack.

Algebra reached Italy in the 13th century. Leonardo of Pisa's *Liber Abbaci* discussed both algebra and arithmetic with Hindu-Arabic numerals. In the following centuries, a lively tradition of arithmetic and algebra teaching developed in Italy. As Italian merchants developed their businesses, they had more and more need of calculation. The Italian "abbacists" tried to meet this need by writing books on arithmetic and algebra. Several of them discussed examples of cubic equations. In some cases, the examples were chosen so that the equations could be solved, or they were constructed from their solutions. In other cases, the authors presented incorrect ways to solve them. None had a complete solution of the general problem.

There was not much real progress on the problem until the work of Scipione del Ferro and Niccolò Fontana, known as Tartaglia ("the Stammerer"), in the first half of the 16th century. Both men discovered how to solve certain cubics, and both kept their solutions secret. At this time, Italian scholars were mostly supported by rich patrons and had to prove their talent by defeating other scholars in public competitions. Knowing how to solve cubic equations allowed them to challenge others with problems that they knew the others could not solve. Thus, this competition system encouraged people to keep secrets.

In 1535 Tartaglia bragged that he could solve cubic equations, but he wouldn't tell anyone how he did it. Scipione del Ferro, who was dead by this time, had passed his own secret on to his student Antonio Maria Fiore. When Fiore heard of Tartaglia's claim, he challenged him to a competition. It turned out that del Ferro knew how to solve equations of the form $x^3 + cx = d$, and that Tartaglia had discovered how to solve $x^3 + bx^2 = d$. When the time for the contest came, Tartaglia presented Fiore with a range of questions on several different parts of mathematics, but each and every one of Fiore's questions boiled down to a cubic of the kind he could solve. Faced with this, Tartaglia managed to find a solution for this kind of equation, too, and won the contest handily when it turned out that Fiore's knowledge didn't extend much beyond cubic equations.

News of Tartaglia's victory eventually reached Girolamo Cardano, one of the most interesting figures of 16th century Italy. Cardano was a doctor, a philosopher, an astrologer, and a mathematician. In each of those fields he came to be well known and respected throughout Europe. In 1552, for example, he was invited to come to Scotland to help treat the Bishop of St. Andrews, who was suffering from serious asthma attacks. He agreed to go and was successful in curing the Bishop, and that solidified his fame.

Cardano's adventures with the cubic equation happened earlier in his life. Having heard of Tartaglia's solution, Cardano contacted him in 1539 to try to convince him to share the secret. Cardano's many pleas and promises of secrecy[2] eventually convinced Tartaglia, who came to Milan to explain his solution to Cardano. Once in possession of a method for solving a couple of cases of the cubic, Cardano attacked the problem of the general equation and, after six years of intense work, managed to solve it completely. His assistant, Lodovico Ferrari, applied the same set of ideas to the general equation of degree 4 (the *quartic*) and managed to find a solution for that, too.

At this point, Cardano knew that he had made a real contribution to mathematics. But how could he publish it without breaking his promise? He found a way. He discovered that del Ferro had found the solution of the crucial case before Tartaglia had. Since he had not promised to keep *del Ferro's* solution secret, he felt that he could publish it, even though it was identical to the one he had learned from Tartaglia. The resulting book was called *Ars Magna*, meaning "The Great Art," that is, algebra. It contains a complete account of how to solve any cubic equation, with geometric justifications of why the methods work. The book also includes Ferrari's solution of the quartic. Written in Latin, the book reached scholars all over Europe. And, of course, it reached Tartaglia.

Tartaglia was furious, but what could he do? The secret was out. He made public the story of Cardano's treachery, but Cardano was on to other things. Instead, Ferrari contacted Tartaglia and challenged him to a competition. Tartaglia felt that Ferrari was an unimportant

---

[2] According to Tartaglia, Cardano said "I swear to you, by God's holy Gospels, and as a true man of honour, not only never to publish your discoveries, if you teach me them, but I also promise you, and I pledge my faith as a true Christian, to note them down in code, so that after my death no one will be able to understand them." See [59], p. 255.

young man, so at first he was not interested in the challenge unless Cardano could be brought in as well. But in 1548 Tartaglia was offered a professorship *on the condition that he defeat Ferrari in a contest.* He agreed, expecting to win easily. Ferrari, however, knew how to solve the general cubic and quartic equations, and Tartaglia had not absorbed that part of the *Ars Magna.* Tartaglia lost, and he remained resentful of Cardano to the end of his life.

This is not yet the end of the story, however. Applying Cardano's method to equations of the form $x^3 = px + q$, one sometimes ended up with expressions that didn't seem to make any sense. For example, for $x^3 = 15x + 4$, Cardano's method gives

$$x = \sqrt[3]{2 + \sqrt{-121}} + \sqrt[3]{2 - \sqrt{-121}}.$$

Normally, one would conclude from the appearance of roots of negative numbers that the equation has no solution. But in this case the equation *does* have a solution, namely $x = 4$.

Cardano noticed this problem before he wrote the *Ars Magna,* and he asked Tartaglia about it. Tartaglia seems to have had no answer; he just suggested that Cardano had simply not understood how to solve such problems. It fell to Rafael Bombelli to resolve the issue. Bombelli began by discussing the equation given above. He then showed, geometrically, that $x^3 = px + q$ always has a positive solution, regardless of the (positive) values of p and q. On the other hand, he showed that, for many values of p and q, solving this equation led to square roots of negative numbers. What Bombelli did at this point was nothing short of brilliant (for his time). He showed that it is possible to work with square roots of negative numbers and still get reasonable answers! (You can find more details about this in Sketch 17.)

With the cubic and quartic solved, the natural next target was the equation of degree 5. That proved to be much more difficult. In fact, it turned out to be impossible to find a formula for solving the general quintic equation. Proving this required a complete change of point of view, which eventually led to the development of abstract algebra.

For a Closer Look: There is a good account of the solution of the cubic equation in [99, Chapter 9]. Al-Khāyammī's algebra has been translated into English; see [104]. Cardano's *Ars Magna* and his autobiography are also available in English as [25] and [26]. They offer a fascinating glimpse of the ways of thought of one of the most brilliant of Renaissance men.

12 A Cheerful Fact
The Pythagorean Theorem

A sk your Average Educated Person what the Pythagorean Theorem says, and the answer you're likely to get is

$$a^2 + b^2 = c^2.$$

If you press a bit about what those letters a, b, and c stand for, you'll often get a blank stare. If you're lucky, your AEP will remember that a and b are supposed to be the lengths of the shorter sides of a right triangle and that c is the length of the longest side, which they probably remember by its "funny" name, the hypotenuse. (One of us and his sons often referred to it as the "hippopotamus," an allusion to a bad pun that shall not be recalled here.) The hypotenuse even makes a cameo appearance in In Gilbert and Sullivan's 1879 comic operetta, *The Pirates of Penzance.*[1]

When we turn to history for the origins of the theorem, we find that they are hard to trace. Greek tradition associates the theorem with Pythagoras, who lived in the fifth century B.C. The problem is that we hear this from authors who wrote many centuries after the time of Pythagoras. By that time Pythagoras was a legendary figure. There is little evidence that he himself was interested in mathematics. It is known, however, that he was the founder of a society, a group for learning and contemplation called the Pythagorean Brotherhood. Later Pythagoreans did get involved in mathematics, but we know very little about how much and in what way.

If we look around for evidence that people knew the theorem, however, we find it in one form or another all over the ancient world — in Mesopotamia, in Egypt, in India, in China, and yes, in Greece. Some of the oldest references are from India, in the *Śulbasūtras*, dating from sometime during the first millennium B.C. There we read that the diagonal of a rectangle "produces as much as is produced individually by the two sides." Similar statements are found in all of the ancient cultures.

---

[1] Late in Act I, the Major General sings:
> I'm very well acquainted, too, with matters mathematical;
> I understand equations, both the simple and quadratical;
> About Binomial Theorem I am teeming with a lot o' news, —
> With many cheerful facts about the square of the hypotenuse!

What we also find in all the ancient cultures are triples of whole numbers that "work" as sides of right triangles. The most famous, of course, is $(3, 4, 5)$. If we make $a = 3$, $b = 4$, and $c = 5$, then

$$a^2 + b^2 = 9 + 16 = 25 = c^2,$$

and this means that a triangle with sides of length 3, 4, and 5 units will automatically be a right triangle. Such triples aren't easy to find, especially when they involve numbers that are a bit larger, but one finds records of them in most ancient cultures. Most historians think that these were created so that mathematics teachers would have examples that would "come out right." Here's an example of the simplest kind of problem one might set:

> One side of a right triangle is 119 meters long, and the hypotenuse is 169 meters long. How long is the other side of the triangle?

Since $169^2 - 119^2 = 14400 = 120^2$, the answer can be found without having to deal with complicated non-integral square roots. The triple $(119, 120, 169)$ comes from an old Babylonian tablet.

Thus, the evidence suggests that the Pythagorean Theorem was actually known by almost all mathematical cultures well before the time of Pythagoras himself, and these cultures also knew how to find triples of whole numbers that would "fit" the theorem. Two competing explanations of this fact have been proposed. One explanation postulates a common discovery, which would then have to have happened in prehistoric times. The other explanation argues that the theorem is so "natural" that it was independently discovered by many different cultures. The second interpretation is supported by the work of cultural historians of mathematics such as Paulus Gerdes, who has shown (see [65], for example) that, by carefully considering patterns and decorations used by African artisans, one can discover the theorem in a fairly natural way.

Of course, discovering (or supposing) the theorem to be true is very different from finding a proof of it. Here, too, the historical situation is unclear. Perhaps the earliest proofs used "square in a square" pictures like Display 1, which is based on an early Chinese source. The idea is to arrange four identical triangles around a square whose side is their hypotenuse. To make it easier to explain what's going on, we have labeled the two shorter sides a and b and the hypotenuse c. Since all four triangles are identical, the inner quadrilateral is a square of side c.

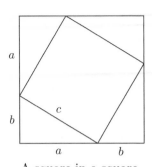

A square in a square

Display 1

How does the diagram prove the theorem? Well, the big square has side $a + b$, so its area is equal to

$$(a + b)^2 = a^2 + b^2 + 2ab.$$

On the other hand, it clearly decomposes into a square with area c^2 and four triangles, each of which has area $\frac{1}{2}ab$. So the decomposition shows that the area of the big square is also equal to $c^2 + 2ab$. Setting these equal and cancelling $2ab$, we see that $a^2 + b^2 = c^2$.

We can also avoid using algebra and make the argument entirely geometric. Rearrange the four triangles as in Display 2. We get the same square of side $a + b$, but now the four triangles have been moved together into two rectangles. The remaining area clearly is two squares, one of side a, the other of side b. So, in the square of Display 1 we have c^2 plus the four triangles, and in the square of Display 2 we have $a^2 + b^2$ plus the four triangles. Hence $a^2 + b^2 = c^2$. In this way, Displays 1 and 2 together make up a "proof without words" of the Pythagorean Theorem.

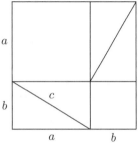

Display 1 rearranged

Display 2

Here is another "proof without words," attributed to Thābit ibn Qurra, a 9th century Islamic mathematician of Baghdad, in what is now Iraq:

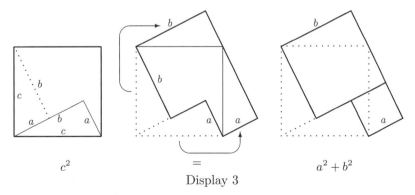

c^2 $=$ $a^2 + b^2$

Display 3

As you might expect, there are many other ways to prove the Pythagorean Theorem. In fact, there are whole books devoted to different proofs of it, many found by amateur mathematicians. Even U.S. President James Garfield (who once said his mind was "unusually clear and vigorous" when studying mathematics) is credited with a proof.

The most famous of all the proofs is the one found in the first book of Euclid's *Elements*. (See Sketch 14.) Euclid's Book I starts off with a list of definitions and basic assumptions. Then follow many propositions (i.e., theorems) about triangles, angles, parallel lines, and parallelograms. The 47th proposition says that

> in right-angled triangles the square on the side opposite the right angle equals the sum of the squares on the sides containing the right angle.

This is a statement about *areas*, not about the lengths of the sides. That is to be expected; in early Greek mathematics, magnitudes were not usually described by numbers. (See page 18 for more on this.)

Display 4 shows the diagram that accompanies the proposition. The triangle is drawn so that the hypotenuse ("the side opposite the right angle") is horizontal, and squares are drawn on each of the sides. The goal is to prove that the big square below is equal to the sum of the two squares above.

Euclid's proof is interesting. He drops a perpendicular from the upper vertex of the right triangle, splitting the bottom square into two pieces. Then, using basic facts about triangles and parallelograms, he proves that each piece of the bottom square is equal to the corresponding smaller square. In other words, he actually shows how to divide the big square into two pieces whose areas match the areas of the two smaller squares.

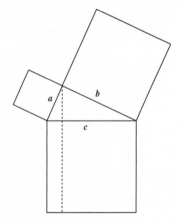

Euclid's proof

Display 4

Euclid goes on, in the next proposition, to prove the converse: "If in a triangle the square on one of the sides equals the sum of the squares on the remaining two sides of the triangle, then the angle contained by the remaining two sides of the triangle is right." This is important, too. It explains, for example, why a $(3, 4, 5)$ triangle can be used to make sure that an angle is a right angle.

Later on in his book, Euclid pushes the theorem one step further: He shows that there is nothing special about "squares" in the theorem. If you draw a geometric figure with its base equal to one of the sides, then draw similar figures with bases equal to the other sides (see Display 5), it is still true that the figure on the hypotenuse is equal (in area) to the sum of the other two figures. This is because the ratio between the areas of similar figures is equal to the ratio of the squares of their sides. So if we draw similar figures with sides equal to a, b, c, their areas are going to equal ka^2, kb^2, and kc^2, where k is a constant.[2] Therefore

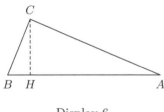

Three similar figures

$$kc^2 = k(a^2 + b^2) = ka^2 + kb^2.$$

Display 5

Moreover, if we can prove this equality for any one value of k, we could cancel k, and we would have a proof of the Pythagorean Theorem. Geometrically, this means that if we could prove the theorem for any specific figure, the general theorem would follow. That leads to what is perhaps the neatest proof of the theorem, which seems only to have been discovered rather recently.

Display 6

Start with the right triangle, and drop a perpendicular from the right angle onto the opposite side. Call the intersection point H. Then it is easy to see that triangles ABC, ACH, and CBH are similar. (Each of the smaller triangles shares an angle with the big one, and all three contain one right angle.) But that means we have constructed three similar figures: triangle ABC on the side AB, triangle ACH on the side AC, and triangle CBH on the side BC. And it is obvious that the two small triangles add up to the big one! So this is enough to prove the theorem.[3]

---

[2]A first figure that is similar to a second can be expanded or shrunk to match that second figure by some (linear) scaling factor. Let k be the area of the figure on a segment of length 1. To get the the figure on a segment of length a, the linear scaling factor is a, so to find its area we need to multiply k by a^2, and similarly for the other two segments.

[3]So why didn't Euclid use this proof, or something like it? Probably because similarity doesn't occur until Book VI, and the Pythagorean Theorem is in Book I.

The Pythagorean Theorem remains tremendously important. It is one of the most useful results in elementary geometry, both theoretically and in practice. For instance, it can be used to make the corners of a garden bed or a garage foundation square: Just mark off 3 feet along one side of a corner and 4 feet along the other, and then adjust their directions until the diagonal between those two marked points is exactly 5 feet. A well known formula of coordinate geometry also follows directly from this theorem: The distance between two points with coordinates (x_1, y_1) and (x_2, y_2) is

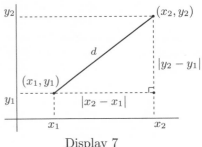

$$d = \sqrt{(x_2 - x_1)^2 + (y_2 - y_1)^2}.$$
Display 7

In fact, from a modern point of view, this distance formula is what makes classical coordinate geometry "Euclidean." If distances were measured some other way, the resulting geometry would not be Euclidean. For example, if those coordinates are latitudes and longitudes of points on the surface of a sphere, their distance is *not* given by that formula. That's because the surface of a sphere is not a plane, and its geometry doesn't obey Euclid's axioms. (See Sketch 19.)

The Pythagorean Theorem is famous because it is useful and because it tells us that the geometry we are studying is Euclidean. But most of all, it is famous because it is beautiful. The relationship it reveals between the sides of a right triangle is unexpected, simple, and... well, right.

For a Closer Look: Many books discuss the history of the Pythagorean Theorem. Chapter 4 of [86], entitled "Many Cheerful Facts About the Square of the Hypotenuse," is a good overview. See also [94] and [65] for information on non-Western cultures, and [48] for a careful discussion of Euclid's proof. For a discussion of "President Garfield's proof," see [89].

13 A Marvelous Proof
Fermat's Last Theorem

Pierre de Fermat was born in 1601 to a moderately well-to-do French family. He went to law school and eventually became a councillor at the parliament of the French city of Toulouse. Rising through the ranks, he finally became a member of the city's criminal court. As a judge, Fermat had a reputation for having a good mind but being somehat distracted. He died in 1665 in Castres, a nearby town.

That quick summary of Fermat's public life does not give any hints about why we still remember and talk about him. The reason, of course, is that there was another side to the man. At some point, probably while he was at the university in Bordeaux, he discovered mathematics. That discovery became his lifelong passion.

Like many scholars of his time, Fermat began his mathematical work by studying the works of Greek mathematicians. One of the first things he did was to "restore" one of the books written by the great geometer Apollonius. A partial record of the contents of this book had survived (mostly a list of results without proofs), and Fermat worked through it, filling in gaps and providing full proofs. Inspired by Greek geometry, he developed several important new ideas. For example, in order to solve certain geometric problems, he invented a method of describing curves by equations, a form of coordinate geometry. (See Sketch 16.) He also developed methods for finding maxima, minima, and tangents, anticipating some of the ideas that Newton and Leibniz would use when they invented calculus late in the 17th century.

Fermat never published any of this work. Instead, he wrote letters describing it, first to friends and eventually to other mathematicians, too. His work on geometry aroused quite a bit of interest, so he went into considerable detail in his letters. As a result, we understand this part of Fermat's mathematics quite well.

But other mathematical questions also fascinated Fermat. These had to do with whole numbers. Perhaps his interest was initially sparked by the centuries-old search for "perfect numbers" — numbers that equal the sum of their proper divisors. In any case, Fermat soon obtained new and interesting results in number theory. For example, he determined which whole numbers could be written as the sum of two squares.

121

Somewhere along the line, he came into contact with the work of Diophantus, another mathematician of Ancient Greece. Diophantus is somewhat unique among the Greek mathematicians whose works have been preserved. His book is about solving problems such as:

> Find three squares such that the difference between the greatest and the middle has a given ratio to the difference between the middle and the least.

The solutions Diophantus looked for were always fractions (ratios of whole numbers) rather than arbitrary real numbers, and this restriction makes his problems quite difficult. His book includes solutions to all the problems, but no real explanations about how such solutions might be found.

Diophantus had a huge impact on Fermat. "Problems about numbers," as Fermat described them, began appearing in his correspondence. Fermat made one big change: Instead of looking for fractions, he wanted whole number solutions. But the problems were similar:

- Show that no cube can be a sum of two cubes.

- Show that the there are infinitely many squares such that if we multiply them by 61 and then add 1, the result is a square.

- Show that every number can be written as the sum of four squares.

- Show that the only whole-number solution of $x^2 + 4 = y^3$ is $x = 11$, $y = 5$.

These problems are quite hard, and most of Fermat's correspondents were unable to solve them. Most of them did not, in fact, seem very interested in this kind of thing. One of the problems was the "negative" nature of some of these results. Mathematicians were supposed to *solve* problems, not to show they were unsolvable! Nevertheless, Fermat kept writing letters about these results, trying to generate interest, with little success. As a result, very few details about this work ever got written down.

In 1670, after Fermat's death, his son Samuel decided to publish some of his father's notes. In particular, he discovered that his father had made many marginal notes in his copy of Diophantus. Rather than publishing just the notes, he prepared a whole new edition of Diophantus which incorporated Fermat's notes. That's where the biggest mystery really starts.

One of Diophantus's problems asked that a given square (such as 25) be written as the sum of two squares (such as $16 + 9$). Next to this problem, Fermat wrote

> *In contrast, it is impossible to divide a cube into two cubes, or a fourth power into fourth powers, or in general any power beyond the square into powers of the same degree. I have discovered a marvelous proof of this, but this margin is too narrow to contain it.*

In other words, Fermat claimed that the equation $x^3 + y^3 = z^3$ has no solution in whole numbers, and similarly for all equations $x^n + y^n = z^n$ for every exponent n bigger than 2. And he added that he could prove this. Many people since have wished his book had bigger margins!

In the 18th century, people began to realize how deep and important Fermat's "problems about numbers" really were. The real mover behind all this was Leonhard Euler, who went through Fermat's number theory, put it in order, and found proofs for most of the claims Fermat had made. He even managed to find one (just one!) case in which Fermat had made a mistake.

That statement in the margin, however, was hard to prove. That there are no integer solutions of $x^4 + y^4 = z^4$ follows from one of the theorems Fermat actually proved: The area of a right triangle whose sides have integer length cannot be a square. Euler managed to find a proof that $x^3 + y^3 = z^3$ has no solutions. But there he got stuck. As he remarked, the two arguments are so different that they give no hint as to how to find a proof for the case $n = 5$, much less a general proof. Because this statement was the only one of Fermat's assertions that remained unproved, it eventually became known as "Fermat's Last Theorem." Of course, to actually be a theorem it needed a proof. Fermat said he had found one, but no one else seemed able to discover that "marvelous proof."

Early in the 19th century, people once again got interested in the problem of trying to prove Fermat's claim. Nobody was able to prove the whole thing, but several mathematicians were able to nibble at the edges of the problem. It quickly became clear that it was enough to find a proof for each prime number value of n. One of the most interesting results was due to Sophie Germain.[1] Germain's approach

---

[1] Sophie Germain discovered mathematics when she was a young girl in Paris. She faced severe opposition from her parents, who mirrored their society's prejudice against female intellectuals and did everything they could to keep Germain away from her studies. She persisted, and eventually her extraordinary talent was recognized by some of the greatest mathematicians of her time. For more about Germain's mathematical work, see p. 52.

to the search for a proof of Fermat's Last Theorem was to split it into two parts. First, one would show that there are no solutions in which the three numbers x, y, and z are not divisible by the exponent n. Then one would attack the "second case," in which one of the three numbers is divisible by n. Then Germain proceeded to show the very first result about the problem that had some generality. She showed that if n is prime and $2n + 1$ is *also* prime, then there are no solutions of $x^n + y^n = z^n$ with none of x, y, or z divisible by n. In other words, she showed that the "first case" was true for any prime exponent n that satisfied the additional constraint that $2n + 1$ should also be prime.

At that time, it was difficult for a woman to publish her mathematical work. As a result, Germain's theorem first appeared (with due credit) in a book by Adrien-Marie Legendre published in 1808. Legendre, who was then in his late fifties, was well known and respected, and his publication of Germain's work helped establish her reputation.

Germain's theorem is quite powerful. Her ideas eventually led to a proof of the "first case" of Fermat's Last Theorem for all exponents up to 100. In particular, if $n = 5$, then $2n + 1 = 11$, which is prime, so that Germain's theorem is enough to prove the first case of Fermat's Last Theorem for $n = 5$. In 1825, Lejeune Dirichlet, then only in his 20s, presented a partial proof of the second case for $n = 5$ to the Paris Academy. Soon after, Legendre finished off the proof, showing that he was still a powerful mathematician even though he was over 70 years old. A few years later, Gabriel Lamé found a way to prove the theorem for $n = 7$. Things seemed to be going well.

Around 1830, Lamé had a brilliant idea. The main difficulty in the equation, he felt, was that on one side there was a sum $x^n + y^n$ and on the other a product z^n. If it were possible to factor $x^n + y^n$, the equation would be much easier to handle. Of course, apart from pulling out a factor of $x + y$, it isn't possible to factor $x^n + y^n \ldots$ unless one uses complex numbers. So Lamé used a complex number ζ such that $\zeta^n = 1$ to factor the expression into a product of terms that were combinations of whole numbers and the complex number ζ. Assuming that these new numbers had the same properties as the ordinary whole numbers (specifically, assuming that such expressions could be factored uniquely as products of "prime" expressions), he sketched out a proof of the full theorem.

Lamé presented his proof to the Paris Academy. Joseph Liouville was suspicious. Why should the new numbers behave like the old ones? Knowing that German mathematician Ernst Kummer had worked with similar ideas, he wrote Kummer asking about this. Kummer replied that he had known for a while that the new numbers *did not* have the required unique factorization property, so Lamé's proof did not work. But Kummer got interested in the problem, and he ended up proving a far-reaching result. He identified a nice property that many prime numbers have, and called such primes "regular." Then he gave a proof of Fermat's Last Theorem for any exponent n which is a regular prime. This doesn't cover all the primes, but it does cover a great many. And Kummer's ideas led to further advances that allowed mathematicians to extend his method to other primes, as well.

Nevertheless, for a long time after Kummer there was little progress towards a *general* proof. In 1909 Paul Wolfskehl, a wealthy German mathematician, established a prize fund of 100,000 marks to reward whoever could find a proof. The prize generated many more failed proofs before it disappeared with the collapse of the German economy after World War I.[2] For

decades after that, the prospect of fame alone was enough to generate many more erroneous attempts, but a real, correct proof still seemed very distant.

It was a surprise, then, when Andrew Wiles announced in 1993 that he had found a proof. What had happened was that, around 1987, Kenneth Ribet had proved a theorem establishing a link between Fermat's Last Theorem and another well-known conjecture that had been suggested in the 1950s and remained unproved. This set Wiles to working, and, after many years of solitary work, he managed to prove enough of the conjecture to prove Fermat's Last Theorem.

Considering that Fermat's marginal note had been written sometime in the 1630s and, centuries later, had become the most famous open problem in mathematics, one can understand the worldwide reaction to the Wiles announcement. E-mail messages flew around. The news was reported on the front page of the *New York Times* and on the NBC Nightly News. Everyone was eager to see the proof.

---

[2]In the 1920s, 100,000 marks would barely buy an ordinary German postage stamp.

Then came several months of silence. Wiles had sent the manuscript to a journal, and the referees were reading it to make sure it was correct. Rumors about difficulties with the proof started to fly. Finally, in December of 1993, Wiles broadcast an e-mail message saying that there was indeed a gap in the proof and that he was working to fix it.

There followed several months of what must have been feverish work. Wiles has said that he found the first six years of working on the problem pleasant, but that the period of trying to fix "the gap" was agony. Nevertheless, he was eventually successful. In September of 1994, he announced that the proof was complete and circulated two manuscripts. One was by Wiles and contained the proof, except for one step for which it referred to the other paper. The other was by Wiles together with his former student Richard Taylor. It contained the crucial step that completed the proof. After 350 years or so, Fermat's Last Theorem was actually a theorem!

For a Closer Look: There are several books about Fermat's Last Theorem, partly because it is so famous and partly because of the drama of Wiles announcement, retraction, and final triumph. The friendliest source of more information, however, is a television program, *The Proof*, which aired on PBS in the *Nova* series of science programs. Also interesting is *Fermat's Last Tango*, a musical play about Wiles's struggle. One of the producers of *The Proof* is the author of [161], which probably hits the best balance between accessibility and correctness among all the books on the subject. The best biography of Fermat is [118] (not an easy book to read). There is more information on Sophie Germain in [114]; see also the historical novel [129].

On Beauty Bare
Euclid's Plane Geometry

"Euclid alone has looked on beauty bare," wrote poet Edna St. Vincent Millay in her *Sonnet xlv.* Why would an artist claim that a mathematician has been the only one to really perceive beauty? One of our goals in writing this sketch is to give you some idea of how to answer that question.

About 2300 years ago in Alexandria, a Greek city near the mouth of the Nile in Egypt, a teacher named Euclid created the world's most famous axiomatic system. His system was studied by Greek and Roman scholars for a thousand years, then translated into Arabic around 800 A.D. and studied by Arab scholars, too. It became the standard for logical thinking throughout medieval Europe. It has been printed in more than 2000 different editions since it first appeared as a typeset book in the 15th century. That system is Euclid's description of plane geometry, and its story really begins at least 300 years before Euclid was born.

According to Greek historians, geometry as a logical discipline began with Thales, a wealthy Greek merchant of the 6th century B.C. They describe him as the first Greek philosopher and the father of geometry as a deductive study. Rather than relying on religion and mythology to explain the natural world, Thales began the search for unifying rational explanations of reality. His search for an underlying unity in geometric ideas led him to investigate logical ways to derive some geometric statements from others. The statements themselves were well known, but the process of linking them with logic was new. The Pythagoreans and other Greek thinkers continued the logical development of geometric principles.

By Euclid's time, the Greeks had developed a lot of mathematics, virtually all of it related to geometry or number theory. The work of Pythagoras and his followers had been around for two centuries, and many other people had written about their own  mathematical discoveries, too. Plato's philosophy and Aristotle's logic were firmly established by then, so scholars knew that mathematical facts should be justified by reason. Many of these mathematical results had been proved from apparently more basic ideas. But even these proofs were disorganized, each one starting from its own assumptions, without much regard for consistency.

Building on this earlier work, Euclid organized and extended a large portion of what the Greek mathematicians had learned. One of his purposes seems to have been to put Greek mathematics on a unified, logical foundation. Euclid set out to rebuild those fields "from the ground up." He wrote an encyclopedic work called the *Elements*, separated into thirteen parts called "books" (each one probably corresponding to a long papyrus scroll):

— Books I, II, III, IV, and VI are about plane geometry;

— Books XI, XII, and XIII are about solid geometry;

— Books V and X are about magnitudes and ratios; and

— Books VII, VIII, and IX are about whole numbers and their ratios.

These books contained a total of 465 "propositions" (we might call them *theorems*), each one proved from statements coming before it. The style of presentation is very formal and dry. There is no discussion or motivation. After the statement of each proposition is a figure to which it refers, followed by a detailed proof. The proofs end with a restatement of the proposition "which was to be proved." The Latin translation of that phrase, "quod erat demonstrandum," is the source for the abbreviation Q.E.D. that still often appears at the end of formal proofs.

Euclid paid special attention to geometry. As Aristotle had already pointed out, a logical system must begin with a few basic assumptions that we take for granted and on which we build. So, after giving a long list of definitions, Euclid specified a small number of basic statements that appeared to capture the essential properties of points, lines, angles, etc., and then he tried to derive the rest of geometry from these basic statements by careful proof. His goal was to systematize the observable relationships among spatial figures, which he, like Plato, Aristotle, and the other Greek philosophers, regarded as ideal representations of physical entities.

For sections that dealt with other topics, Euclid followed the same procedure, making new definitions and new assumptions and then building the theory on those assumptions. Book V is particularly important: It contains a detailed theory of ratios among quantities of various types. These ratios played a crucial role in Greek mathematics, and the foundation provided in this book (which tradition says is due to Eudoxus) was therefore very important.

With a touch of genius, Euclid connected his entire work to Plato's philosophy. In the last book of the *Elements*, he proved that the only possible types of regular polyhedra[1] are the five Platonic Solids, which symbolized for Plato the basic elements of the entire universe. (See Sketch 15.)

Book I begins with ten basic assumptions:[2]

COMMON NOTIONS

1. Things equal to the same thing are also equal to each other.

2. If equals are added to equals, the results are equal.

3. If equals are subtracted from equals, the remainders are equal.

4. Things that coincide with one another are equal to one another.

5. The whole is greater than the part.

POSTULATES

1. A straight line can be drawn from any point to any point.

2. A finite straight line can be extended continuously in a straight line.

3. A circle can be formed with any center and distance (radius).

4. All right angles are equal to one another.

5. If a straight line falling on two straight lines makes the sum of the interior angles on the same side less than two right angles, then the two straight lines, if extended indefinitely, meet on that side on which the angle sum is less than the two right angles.

In modern terminology, all ten of these starting-point statements are the *axioms* for Euclid's plane geometry. The first five are general statements about quantities that Euclid clearly considered to be obviously true. The second five are specifically geometric. In Euclid's view, these five statements are intuitively true. In other words, anyone who knows what the words mean will believe them. To clarify the meanings of the words, he provided 23 definitions or descriptions of the basic terms of geometry, starting with *point* and *line*.

(Does the fifth postulate seem odd to you? It appears to be true, but its language is much more complicated than the others. Many

---

[1] *Regular polyhedra* are three-dimensional shapes made up entirely of congruent polygonal faces.

[2] These and other statements of Euclid are adapted from [49].

mathematicians throughout history were bothered by that. For the full story of where this led, see Sketch 19.)

From this simple beginning — twenty-three definitions, five Common Notions, and five Postulates — Euclid reconstructed the entire theory of plane geometry. His work was so comprehensive and clear that the *Elements* became the universally accepted source for the study of plane geometry from his time on. Even the geometry studied in high school today is essentially an adaptation of Euclid's *Elements*.

The enduring importance of Euclid's work stems from one simple fact:

> The *Elements* is not just about shapes and numbers; it's about how to think!

Not just about mathematics. Euclid shows you how to think logically about anything — how to build a complex theory one step at a time, with each new piece firmly attached to what has already been built. Euclidean plane geometry has shaped Western thought over the years. In fact, many of the most influential writings in politics, literature, and philosophy cannot truly be understood without some appreciation of Euclid. For example:

- In the 17th century, French philosopher René Descartes based part of his philosophical method on the "long chains of reasoning" used in Euclid to move from simple first principles to complex conclusions.

- Also in the 17th century, British scientist Isaac Newton and Dutch philosopher Baruch Spinoza used the form of Euclid's *Elements* to present their ideas.

- In the 19th century, Abraham Lincoln carried a copy of Euclid with him and studied it at night by candlelight in order to become a better lawyer.

- On July 4, 1776, the 13 American colonies broke away from Great Britain by agreeing to an axiomatic system, the Declaration of Independence. After a brief opening paragraph, the axioms are explicitly stated as self-evident truths. The document goes on to prove a fundamental theorem: The 13 American colonies are justified in breaking away from Great Britain and forming an independent country — the United States of America.

We hold these truths to be self-evident:
– that all men are created equal;
– that they are endowed by their Creator
with certain unalienable rights;
– that among these are life, liberty, and
the pursuit of happiness.
That, to secure these rights...

It is exactly as a model of precise thought that Euclid was studied for many centuries. Students would either work through the *Elements* itself or study some simplified or "improved" version. Most of them did not get very far. In fact, one theorem early in Book I became known as the *pons asinorum* ("the bridge of asses") because it was the place where weaker students began to have difficulties. Not all students enjoyed the exercise. At Yale in the 19th century, students developed an elaborate ritual at the end of their sophomore year to celebrate the fact that their mathematical studies were complete. It was called the Burial of Euclid. At one point in the ritual,

> Euclid's volume was perforated with a glowing poker, each man of the class thrusting the iron through in turn to signify that he had gone through Euclid. Following this the book was held for a moment over each man to betoken that he had understood Euclid, and finally each man passed the pages under foot that he might say thereafter that he had gone over Euclid.[3]

This was followed by a funeral cortege, a funeral oration, and the cremation of the book!

In the 20th century, the study of Euclidean geometry migrated from the universities to the high schools. The "two-column proof," in which each step in the left column must be justified by a reason in the right column, seems to have been invented early in the 1900s as a way to make it easier for students to understand and construct proofs. However, its rigid structure led far too often to a student strategy of "learning" proofs by rote, memorizing the steps without understanding either the logic of the argument or the significance of the theorem. As a result, many students regarded high school geometry as a painful, irrelevant ritual with no connection to their "real world."

---

[3]See [113], pp. 78–79.

From about the 1970s on, high school geometry texts began to compensate for this by inserting various other ideas and approaches into the course, including more and more informal geometry, discussions of measurement, and the like. Unfortunately, these well-meaning attempts to make this single course "serve two masters" often ended up doing justice to neither and blurring the focus of what was being studied. Gradually many of these courses have become almost entirely informal, emphasizing student "discovery" of geometric ideas via group activities and discussion. Euclid's logical structure is relegated to a final chapter or two, if it appears at all, making it a likely candidate for omission by teachers who find themselves pressed for time.

This de-emphasis of Euclid's logical structure in high school geometry is truly unfortunate. In today's world, the ability to view a situation in axiomatic terms and to deal with its logical structure is still very important, and not just in mathematics. For instance, it is immensely helpful in understanding, negotiating, and enforcing collective bargaining agreements, which govern the working conditions of a large part of the U.S. workforce; in dealing with computer systems, software packages, and the like, which are rapidly becoming a central part of everyday life; and in coping intelligently with the arguments that swirl around the hot social-political-legal issues of the day, such as abortion, GLBT rights, affirmative action, and equal opportunity.

Mathematician E. T. Bell once said "Euclid taught me that without assumptions there is no proof. Therefore, in any argument, examine the assumptions." The prototypical logical system underlying all such axiomatic analysis is the plane geometry of Euclid, a way of organizing ideas that is as relevant today as it was when Euclid first wrote it down, some 2300 years ago.

For a Closer Look: The best English translation of Euclid's *Elements* is Sir Thomas L. Heath's *The Thirteen Books of the Euclid's Elements* [49]. Heath's commentary is outdated and should not be trusted, so [50], a one-volume edition of Heath's text without the notes, is a better choice. There is also an online edition by David Joyce with the diagrams done as Java applets. Chapters 2 and 3 of [48] contain an accessible discussion of some of Euclid's theorems, with lots of historical context, as do several books on geometry, such as [86], [52], and [141]. Finally, [7] gives a good survey of all of Euclid's *Elements*, including the nongeometric parts.

In Perfect Shape
The Platonic Solids

The Greeks were very fond of symmetry. You can see it in their art, in their architecture, and in their mathematics. In plane geometry, a major area of Greek mathematics, the most symmetric polygons are the *regular* ones — polygons with all sides and all angles congruent. A regular triangle is one that is equilateral; a regular quadrilateral is a square.

In three-dimensional space, a polyhedron is *regular* if all its faces are congruent regular polygons and all its vertices are similar. For example, a cube is a regular polyhedron; all its faces are squares of the same size and each vertex has four squares adjacent to it. It is a remarkable fact of geometry, proved as the final proposition of Euclid's *Elements*, that only a few convex regular polyhedra exist. (Contrast this with regular polygons, which can have any number of sides.) In fact, there are exactly five different types, as pictured in Display 1.

tetrahedron — 4 faces (triangles)
hexahedron (cube) — 6 faces (squares)
octahedron — 8 faces (triangles)
dodecahedron — 12 faces (pentagons)
icosahedron — 20 faces (triangles)

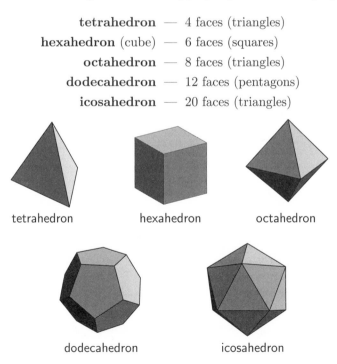

tetrahedron   hexahedron   octahedron

dodecahedron   icosahedron

Display 1

The fact that there are only these five regular polyhedra may seem puzzling at first, but it is not hard to see why that must be so. Think of it this way:

- To form some sort of point or "peak," at least three polygonal faces must meet at any vertex of the polyhedron.

- Since the polyhedron is regular, the situation at any vertex is the same as at any other. Therefore, we only have to consider what happens at a typical vertex.

- In order to make a peak, the sum of all the face angles at the vertex must be less than 360°. (If they added up to exactly 360°, they would make a flat surface.)

- Since all the faces are congruent, the angle sum at a vertex must be divided up equally among them.

- Now let's look at the possible types of regular polygons that might be used as faces of these polyhedra:

Triangles: Each angle of an equilateral triangle measures 60°. How many could you have together and still total less than 360°? 3 (180°), 4 (240°), or 5 (300°). (See Display 2.) That's all. Putting six together would give you a flat "peak," and more than that would be too large a total. These three possible cases give you the tetrahedron, octahedron, and icosahedron.

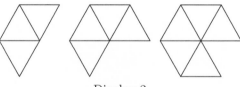

Display 2

Squares: Each angle measures 90°. Three of them (totaling 270°) could meet at a vertex (of a cube), but four is too many. (See Display 3.)

Display 3

Pentagons: Each angle of a regular pentagon measures 108°. Three of them (totaling 324°) could meet at a vertex (of a dodecahedron), but four is too many. (See Display 4.)

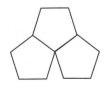

Display 4

Hexagons: Each angle of a regular hexagon measures 120°. Three of them total 360°, which is too much! Thus, there is *no* regular polyhedron with hexagonal faces.

Other regular polygons: Each angle of a regular polygon with more than six sides must measure more than 120°. Three of them coming together at a point would total more than 360°. Therefore, there is *no* regular polyhedron with faces of any other kind.

The regularity of these shapes implies that each one can be *inscribed* in a sphere; that is, it can be placed inside a spherical shell in such a way that all its vertices are on the sphere. This was of great significance to the Pythagoreans, who connected four of the polyhedra with the four "elements" of the physical world:

$$
\begin{array}{rcl}
\text{fire} & - & \text{tetrahedron} \\
\text{earth} & - & \text{hexahedron (cube)} \\
\text{air} & - & \text{octahedron} \\
\text{water} & - & \text{icosahedron}
\end{array}
$$

Plato discusses the connection between the elements and the polyhedra in one of his dialogues, the *Timaeus*, which tells stories about the creation of the cosmos. In it, the Creator is said to have used the tetrahedron, octahedron, icosahedron and cube as models for the fundamental particles of fire, air, water, and earth. The problem was what to do with the dodecahedron. (It had to have *some* significance!) Plato says that it served as a model for the whole universe, and so was a "fifth element," the fundamental essence of the universe. This is where we get the word "quintessence," which today means the best, purest, and most typical example of some quality, class of persons, or nonmaterial thing. Because of Plato's interest in them, these five polyhedra are known as "the Platonic Solids" or "the Platonic Bodies."

Can we get more solids if we relax the demands of regularity? Suppose we require that all the faces be regular polygons, but not necessarily all the same kind. So, for example, the faces could be either triangles or pentagons. We'll still require that all the triangles be congruent, that all the pentagons be congruent, and that all the vertices be similar.

The Greek mathematician Pappus tells us that Archimedes had considered this possibility and discovered that there were exactly 13 such solids. (Because of this, they are sometimes called the *semi-regular polyhedra* or the "Archimedean Solids.") Here's an example. Start with an icosahedron. It has triangular faces which come together, five at a time, in 12 vertices. Suppose we cut off each of these peaks.

This will replace each of the 12 vertices by a new face, which will be a pentagon. If we do the cutting carefully, we can arrange that the surviving part of each of the 20 triangular faces is a regular hexagon. So we'll end up with a solid that has 20 hexagonal faces and 12 pentagonal ones. It is usually called the truncated icosahedron, and it actually exists in the "real world": it is the pattern on the cover of a traditional soccer ball.

In the Renaissance, mathematicians once again became fascinated by the regular and semi-regular solids. They learned about the five regular solids from Plato and Euclid, but most of them never read Pappus, so they had to rediscover the Archimedean solids. They did so slowly, with great excitement. This work culminated with Johannes Kepler (better known for his work in astronomy), who found all thirteen Archimedean solids and proved that there are no others.

Kepler, like Plato, tried to relate these beautifully symmetric solids to the real world. He once attempted to construct a theory of the solar system based on the Platonic Solids. He imagined that the orbit of each planet was in a large sphere. Then he said that if one inscribed a cube into the sphere of Saturn, the faces of the cube would be tangent to the sphere of Jupiter. Similarly, inscribing a tetrahedron in the sphere of Jupiter made the faces tangent to the sphere of Mars. And so on with the dodecahedron, the icosahedron, and the octahedron. Luckily, Kepler eventually gave up on this idea and went on to discover that the orbits of the planets are actually ellipses.

Plato's triangular theory of matter and Kepler's polyhedral theory of the solar system have not stood the test of time, but the Platonic solids can still be found in the earth's elements:

— the crystalline structures of lead ore and rock salt are hexahedral;

— fluorite forms octahedral crystals;

— garnet forms dodecahedral crystals;

— iron pyrite crystals come in all three of these forms;

— the basic crystalline form of the silicates (which form about 95% of the rocks in the Earth's crust) is the smallest of the regular triangular solids, the tetrahedron; and

— the sixty carbon atoms in the molecule known as the "buckyball" are arranged at the vertices of a truncated icosahedron.

For a Closer Look: See [33] for a full-length discussion of polyhedra that includes extensive historical information.

Shapes by the Numbers

Coordinate Geometry

16

One of the most powerful ideas in all of mathematics is the understanding of how to represent shapes by equations, a field we now call *analytic geometry*. Without this bridge between geometry and algebra there would be no calculus for science, no CAT scans for medicine, no automated machine tools for industry, no computer graphics for art and entertainment. Many things we take for granted simply wouldn't exist. Where did this marvelous insight come from? Whose idea was it, and when did it happen?

When you think of analytic geometry, what's the first thing that comes to mind? For most people, it's a pair of coordinate lines, an *x*-axis and a *y*-axis at right angles to each other, often called a *Cartesian coordinate system*. "Cartesian" refers to the 17th century French philosopher/mathematician René Descartes, who is usually credited with the invention of analytic geometry. He did, indeed, formulate most of the key ideas of analytic geometry, but the rectangular coordinate system as we know it today was not one of them.

In a way, the story begins when surveyors in ancient Egypt used a rectangular grid to divide the land into districts, much the same as our modern road maps are divided into squares for indexing purposes. This enabled them to catalog locations by using two numbers, one for a row and the other for a column. That method was also used by Roman surveyors and Greek mapmakers in early times. However, labeling locations by a number grid is only a small step in bridging the gap between geometry and algebra. The much more fundamental question is the connection between algebraic expressions — that is, equations and functions — and shapes in a plane or in space.

A glimmer of this idea is traceable back to ancient Greece. About 350 B.C. Menaechmus, related some kinds of curves we now call *conic sections* (the curves formed by cutting a cone by a plane, discussed in more detail in Sketch 28) to the solution of numerical proportions. This laid the groundwork for the exploration of conic sections by Apollonius a century or so later. Apollonius and other mathematicians of his

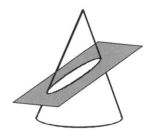

a conic section

day were interested in *locus*[1] questions: What points satisfy a given set of conditions, and do they form some kind of line or curve? For instance, the locus of points a fixed distance from a given point is a circle; the locus of points whose distance from a given point equals the perpendicular distance to a given line is a parabola. Apollonius investigated much more complex locus questions and showed that some (but not all) of them resulted in conic sections. Was this analytic geometry? Not exactly. Apollonius was headed in that direction, but his geometric figures were connected with numerical relationships by means of ratios and words. Development of the pattern-rich symbolic language of algebra was still many centuries in the future.

The evolution of an efficient algebraic symbolism took a long time. (See Sketch 8.) It was not yet in place in the 14th century when Nicole Oresme described a way of graphing the relationship between an independent variable and a dependent one. Late in the 16th century, François Viète attempted to distill the essence of the ancient Greeks' geometric analysis by representing quantities with letters and relationships with equations. In so doing, he took a giant step toward focusing the power of algebra on the problems of geometry. All that was needed at this point was the right creative insight.

That insight came independently and almost simultaneously from two Frenchmen in the first half of the 17th century. One was Pierre de Fermat, perhaps the world's finest mathematical amateur. Fermat was an unobtrusive lawyer and civil servant who indulged in mathematics of all sorts just for the fun of it. He published very little, preferring to exhibit his extraordinary creativity in correspondence with some of the other leading mathematicians of his day.[2] One of those letters tells us shows that Fermat had developed many of the key concepts of analytic geometry by about 1630. Motivated by his interest in the locus problems of Apollonius, Fermat devised a coordinate system of sorts for plotting the relationship between two unknown (positive) quantities, A and E, as follows.

Starting at some convenient point, he drew a horizontal reference line to the right. He marked off one variable, A, from the starting point on the horizontal axis, and at its other end placed a segment at a fixed angle to the first, representing the other variable, E. He envisioned the length of E varying as the length of A varied, according to the equation he was studying, and the position of the E segment moving to

---

[1] *Locus*, the root of our English words *local* and *location*, means "place" in Latin. Its plural is *loci*.

[2] For more about Fermat, see Sketch 13.

the right, always making the same angle with the A segment. Fermat said[3]
that "It is desirable, in order to aid
the concept of an equation, to let the
two unknown magnitudes form an angle, which usually we would suppose
to be a right angle," but in fact he
allowed other angles also.

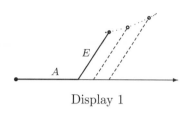

Display 1

Fermat's crucial insight was to see that whenever he could find an
equation relating his two variables A and E, he would get a curve that
corresponded to this equation. The top endpoint of the E segment
would trace out a curve, and it would be the locus associated to the
equation. In Fermat's own words:

> Whenever two unknown magnitudes appear in a final equa-
> tion, we have a locus, the extremity of one of the unknown
> magnitudes describing a straight line or a curve.

(To illustrate the general case, the angle in Display 1 is not a right
angle. If it were, this picture would look a lot like the familiar way of
graphing E as a function of A.)

Notice, in particular, Fermat's claim that this process will give you
a locus for *any* equation. Since it is easy to write down equations, we
suddenly have a huge number of curves available, without having to
create them via some sort of geometric construction. In addition, we
have a natural question: Given a class of equations, describe the corresponding class of curves. Fermat did this for all quadratic equations,
showing, in fact, that the resulting loci are always conic sections.

Fermat's writings on the algebraic approach to geometry were not
published until after his death, so much of the credit for inventing
analytic geometry was given to someone else. The "someone else" is
René Descartes, a French nobleman by birth, a student of mathematics
in his youth, a soldier by choice in his physical prime, and an eminent,
free-thinking philosopher and mathematician in the last twenty years
of his life. Many mathematicians regard the development of analytic
geometry as one of his crowning achievements, but to Descartes it was
just one of three case studies in a much grander scheme. He wrote it
as one of three appendices to a philosophical tract commonly known
as *Discourse on Method*. Its full title clearly reveals his intentions:
*Discourse on the Method of Rightly Conducting Reason and Seeking
Truth in the Sciences*. His actual goal was to redefine the methodology

---

[3]This and the following quote are from Fermat's *Introduction to Plane and Solid
Loci*, in [165], pp. 389–396.

for seeking truth in *all* matters; the word "sciences" had a much broader connotation at that time than it does now. In his own words:

> The long chains of simple and easy reasonings by means of which geometers are accustomed to reach the conclusions of their most difficult demonstrations, had led me to imagine that all things, to the knowledge of which man is competent, are mutually connected in the same way, and that there is nothing so far removed from us as to be beyond our reach, or so hidden that we cannot discover it, provided only we abstain from accepting the false for the true, and always preserve in our thoughts the order necessary for the deduction of one truth from another.[4]

Descartes regarded his method as a synthesis of classical logic, algebra, and geometry, in which he discarded their excesses and fixed their limitations. He sets the stage for this blending of ideas by saying,

> [A]s for Logic, its syllogisms and the majority of its other precepts are of avail rather in the communication of what we already know... than in the investigation of the unknown....
> [A]s to the Analysis of the ancients and the Algebra of the moderns,... the former is so exclusively restricted to the consideration of figures, that it can exercise the Understanding only on condition of greatly fatiguing the Imagination; and, in the latter, there is so complete a subjection to certain rules and formulas, that there results an art full of confusion and obscurity calculated to embarrass, instead of a science fitted to cultivate the mind.

To illustrate the power of his new way of thinking, Descartes wrote three appendices to the *Discourse on Method* — on optics, meteorology, and geometry. In the geometry appendix, simply called *La Géométrie*, were the main ingredients for analytic geometry. His main graphical device was essentially the same as the one devised by Fermat: the independent variable, now called x, marked off along a horizontal reference line, and the dependent variable, now y, represented by a line segment making a fixed angle with the x segment. Descartes, perhaps even more than Fermat, emphasized that the angle choice was a matter of convenience, and need not always be a right angle.

---

[4]This and other quotes from the *Discourse on Method*, are taken from the translation available online at Project Gutenberg.

Besides introducing an improved system of algebraic notation in this appendix (see Sketch 8), Descartes explicitly broke with the Greek-inspired view of powers as geometric dimensions by defining a unit of length and then interpreting all quantities in terms of that unit. In particular, just as x was a *length* relative to that unit, so were x^2, x^3, and any higher power of any unknown. Since his goal was to match geometry and algebra, he also showed how to *construct* segments with those lengths. Since any power of the unkown can be represented as a length, we no longer need to connect squaring to squares and higher powers begin to make sense. This conceptual shift, which was only implicit in Fermat's work, let him (and us) consider curves defined by functions containing various powers of an unknown, graphing them without any associated restriction of geometric dimension. Descartes used his new techniques to solve a major locus problem posed by Apollonius but still unsolved by classical Greek methods, thereby claiming to demonstrate the superiority of his "Method of Rightly Conducting Reason and Seeking Truth."

Despite the power of his algebraic approach to geometry, Descartes's *La Géométrie* did not have an immediate impact on the mathematics of his time. There were several reasons for this. One was the suspicion of some mathematicians that algebra lacked the same level of rigor as classical geometry. Then there was the simple matter of language. Descartes, who was living in Holland at the time, wrote the *Discourse on Method* and its appendices in his native French. However, the "universal" language of 17th-century European scholarship was Latin. Many scholars couldn't read French at all. Also, Descartes had deliberately omitted the details of many of his proofs, saying to his readers that he did not want to "deprive you of the pleasure of mastering it yourself."[5] For example, you won't find the equation of a straight line discussed in *La Géometrie*.

Credit for removing these roadblocks and making the ideas widely available belongs primarily to the Dutch mathematician Frans van Schooten, who translated *La Géométrie* into Latin and published a greatly expanded version, which included a wealth of detailed commentary. During the years 1649 to 1693, Van Schooten's work was published in four editions and grew to be about eight times as long as the original. It was from these Latin versions that Isaac Newton learned about analytic geometry on his way to developing the fundamental ideas of calculus. By the end of the 17th century, the methods of Cartesian geometry were fairly widely known throughout Europe.

---

[5]See [42], p. 10.

But the coordinate geometry of that time did not yet include one feature that we take for granted today — the *ordinate*, or vertical axis! For both Descartes and Fermat, the first coordinate of a point was a line segment starting at the origin and extending to the right, and the second coordinate was a segment that started at the end of the x

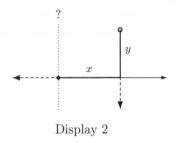

Display 2

segment and extended upward at a fixed angle to the point. It is true that Fermat usually considered this angle to be a right angle (see the quote above), but neither he nor Descartes ever referred to a second axis. Moreover, since they regarded these "unknown magnitudes" to be (lengths of) line segments, they only considered positive coordinates. John Wallis, an English mathematician writing in the 1650s, extended these ideas to include negative coordinates. Yet, as late as the middle of the 18th century, even such a prominent mathematician as Leonhard Euler did not consistently include a vertical axis in coordinate diagrams. The two-axis rectangular coordinate system that we commonly attribute to Descartes seems instead to have evolved gradually during the century and a half after he published *La Géométrie*.

Coordinate (or analytic) geometry is a prominent link in the historical chain of big mathematical advances. Just as the development of symbolic algebra paved the way for analytic geometry, so analytic geometry, in turn, paved the way for calculus. Calculus, in its turn, opened the door to modern physics and many other areas of science and technology. In the past several decades this algebraic way of describing shapes has also combined synergistically with the calculating speed of modern computers to produce increasingly astounding visual images for a wide range of uses. All of this rests on the truly simple idea of giving each point in space a numerical address, so that we can describe shapes by the numbers.

For a Closer Look: The birth of coordinate geometry is discussed in all of the standard references. For a more detailed account, see [17]. For the adventurous, there is [42], a bilingual edition of the original *La Géométrie*.

17 Impossible, Imaginary, Useful Complex Numbers

The standard way of introducing the complex numbers is to argue that we want to be able to solve all quadratic equations, including $x^2 + 1 = 0$. The obvious reaction to that is: "Why would I want to solve that?" It's a good question.

During many centuries of the study of algebraic equations, mathematicians thought of them as a means for solving concrete problems. For "a square and ten things make thirty-nine," the square was pictured as a geometric figure, and the "things" were its sides. (See Sketch 10.) In this context, even negative solutions didn't make much sense. And if applying the quadratic formula led you to the square root of a negative number, this meant that your problem had no solution.

A good example of this can be found in Girolamo Cardano's *The Great Art*, published in 1545.[1] He discusses the problem of finding two numbers whose sum is 10 and whose product is 40. He observes, correctly, that no such numbers exist. Then he points out that the quadratic formula leads to the numbers $5 + \sqrt{-15}$ and $5 - \sqrt{-15}$. He sees that, if he suppresses his disgust at such nonsense and computes with these expressions, he can show that their sum is indeed 10 and that their product is indeed 40. But he dismisses this kind of thing as a meaningless intellectual game. In another book, he says that "$\sqrt{9}$ is either $+3$ or -3, for a plus [times a plus] or a minus times a minus yields a plus. Therefore $\sqrt{-9}$ is neither $+3$ or -3 but is some recondite third sort of thing."[2]

Display 1

As various of his near-contemporaries noted, he had a point. For example, early in the 17th century, René Descartes pointed out that when one tries to find the intersection point of a circle and a line, one has to solve a quadratic equation. The quadratic formula leads to the square root of a negative number exactly when the line does not, in fact, intersect the circle, as in Display 1. So, for the most part, the feeling was that the appearance of "impossible" or "imaginary" solutions was simply a signal that the problem in question did not have any solutions.

---

[1] See Sketch 11 for more on Cardano.

[2] From *Ars Magna Arithmeticae*, problem 38, quoted in [25], p. 220, note 6.

Even in Cardano's time, however, there were indications that life (mathematical life, at least) was not so simple. Cardano's greatest mathematical triumph was finding the formula for solving a cubic equation (see Sketch 11 for the story). For an equation[3] of the form $x^3 + px + q = 0$, Cardano's formula for the solution, recast in modern notation, was

$$x = \sqrt[3]{-\frac{q}{2} + \sqrt{\frac{q^2}{4} + \frac{p^3}{27}}} + \sqrt[3]{-\frac{q}{2} - \sqrt{\frac{q^2}{4} + \frac{p^3}{27}}}.$$

This gave a solution for many cubics, but in some cases there was a glitch. Suppose, for example, that the equation is $x^3 = 15x + 4$. We rewrite it as $x^3 - 15x - 4 = 0$ and apply the formula to get

$$x = \sqrt[3]{2 + \sqrt{-121}} + \sqrt[3]{2 - \sqrt{-121}}.$$

Based on our experience with quadratics, the correct conclusion would seem to be that there is no solution. But if we try $x = 4$ we see that leaping to conclusions is a mistake: the equation does have a real root. (In fact, it has three real roots.)

Cardano noticed this problem, but seems not to have known what to do about it. He mentioned it twice in his book. The first time, he said that this case needs to be solved using a different method to be described in another book. (In a later edition, he referred instead to a chapter describing tricks that might be used to solve certain equations.) The second time, he wrote[4]: "Solving $y^3 = 8y + 3$, according to the preceding rule, I obtain 3." That must have puzzled any reader who tried to work it out "according to the rule," since the computation involves $\sqrt{-1805/108}$.

It was Rafael Bombelli who, in the 1560s, first proposed a way out of the quandary. He argued that one could just operate with this "new kind of radical." To talk about the square root of a negative number, he invented a strange new language. Instead of talking about $2 + \sqrt{-121}$ as "two plus square root of minus 121," he said "two plus of minus square root of 121,"[5] so that "plus of minus" became code for adding a square root of a negative number. Of course, subtracting such a square root became "minus of minus." Since $2 + \sqrt{-121} = 2 + 11\sqrt{-1}$, he also

---

[3]We give equations and formulas in modern notation; see Sketch 8 for the story of algebraic symbolism.

[4]See [25], p. 106

[5]The Italian is *piu di meno*.

referred to this as "two plus of minus 11." And he explained rules of operation such as

> plus of minus times plus of minus makes minus;
> minus of minus times minus of minus makes minus; and
> plus of minus times minus of minus makes plus.

It's natural for us to read those as saying that

> i times i is -1;
> $-i$ times $-i$ is -1; and
> i times $-i$ is 1.

But we should be more careful. Bombelli was not really thinking of these "new kinds of radicals" as numbers. Rather, he seems to have been proposing formal rules that allowed him to transform a complicated expression such as

$$\sqrt[3]{2 + \sqrt{-121}} + \sqrt[3]{2 - \sqrt{-121}}$$

into simpler expressions. He showed that his formal rules led to

$$(2 \pm \sqrt{-1})^3 = 2 \pm \sqrt{-121},$$

so

$$\sqrt[3]{2 + \sqrt{-121}} + \sqrt[3]{2 - \sqrt{-121}} = (2 + \sqrt{-1}) + (2 - \sqrt{-1}) = 4,$$

which is the solution of the cubic that started us down this path. Bombelli doesn't bother to look for any other solutions.

Bombelli's work showed that sometimes the square roots of negative numbers are needed in order to find *real* solutions. In other words, he showed that the appearance of such expressions did not always signal that a problem was not solvable. This was the first sign that complex numbers could actually be useful mathematical tools.

But the old prejudice persisted. A half century later, both Albert Girard and Descartes seem to have known that an equation of degree n will have n roots, provided that one allows for "true" (real positive) roots, "false" (real negative) roots, and "imaginary" (complex) roots. This helped to make the general theory of equations simpler and tidier, but the complex roots were still often described as "sophistic," "impossible," "imaginary," and "useless."[6]

---

[6]For an even longer list of such descriptions, see [76], section 9.3.

The next step seems to have come with the work of Abraham De Moivre in the early 18th century. If you look at the formula for multiplying two complex numbers,

$$(a + ib)(c + id) = (ac - bd) + i(bc + da),$$

with just the right frame of mind, you may notice the similarity between the real part and the formula

$$\cos(x + y) = \cos(x)\cos(y) - \sin(x)\sin(y).$$

The two cosines come together, as do the two real parts of the factors above, and so do the two sines and the two imaginary parts. The imaginary part of the product brings to mind

$$\sin(x + y) = \sin(x)\cos(y) + \sin(y)\cos(x)$$

because sines and cosines are mixed, as are the real and imaginary parts of the factors. From there, it's not hard to get to De Moivre's famous formula:

$$(\cos(x) + i\sin(x))^n = \cos(nx) + i\sin(nx).$$

(This is implicit in De Moivre's work even though it isn't stated there in this form.) A few years later, Leonhard Euler brought all the threads together when he discovered (using calculus) that

$$e^{ix} = \cos(x) + i\sin(x)$$

when x is measured in radians. With $x = \pi$ in that expression we get

$$e^{i\pi} = -1 \qquad \text{or} \qquad e^{i\pi} + 1 = 0,$$

a famous formula that relates some of the most important numbers in mathematics.

By the middle of the 18th century, then, it was known that complex numbers were sometimes necessary steps towards the solution of problems about real numbers. It was known that they played a role in the theory of equations, and it was known that there was a deep connection between complex numbers, the trigonometric functions, and exponentials.

But there were still lots of problems. For example, Euler still got all tangled up in expressions like $\sqrt{-2}$. A real radical has a definite meaning: $\sqrt{2}$ means the *positive* square root of two. But, because complex numbers are neither positive nor negative, there is no good way to choose which square root we mean. So one finds Euler saying that

$$\sqrt{-2} \cdot \sqrt{-2} = -2 \qquad \text{and} \qquad \sqrt{-3} \cdot \sqrt{-2} = \sqrt{(-3) \cdot (-2)} = \sqrt{6}$$

without noticing that if he applied the second method to the first equation he would get the incorrect result

$$\sqrt{-2} \cdot \sqrt{-2} = \sqrt{(-2) \cdot (-2)} = \sqrt{4} = 2.$$

Reasoning by analogy often works, but not always!

Although Euler used complex numbers a lot, he didn't resolve the issue of what they actually were. In his *Elements of Algebra*, he says

> Since all the numbers which it is possible to conceive are either greater or less than 0, or are 0 itself, it is evident that we cannot rank the square root of a negative number amongst possible numbers, and we must therefore say that it is an impossible quantity. In this manner we are led to the idea of numbers, which from their nature are impossible; and therefore they are usually called *imaginary quantities*, because they exist merely in the imagination.
>
> All such expressions as $\sqrt{-1}$, $\sqrt{-2}$, $\sqrt{-3}$, $\sqrt{-4}$, &c. are consequently impossible, or imaginary numbers, ... and of such numbers we may truly assert that they are neither nothing, nor greater than nothing, nor less than nothing; which necessarily constitutes them imaginary, or impossible.
>
> But notwithstanding this, these numbers present themselves to the mind; they exist in our imagination, and we still have a sufficient idea of them... [7]

This attitude typified that of most 18th-century mathematicians: complex numbers were basically useful fictions. Bishop George Berkeley[8] likely would have retorted that *all* numbers are useful fictions; however, he was pretty much the only one to think along those lines at the time.

In the 19th century things began to be sorted out. R. Argand was one of the first one to suggest, in a booklet published in 1806, that one could dispel some of the mystery of these "fictitious" or "monstrous" imaginary numbers by representing them geometrically on a plane: the line segment from $(0,0)$ to (x,y) corresponds to the complex number $x+iy$. Adding two complex numbers corresponded to the parallelogram law for adding vectors, and multiplication corresponded to a "scale and rotate" operation.

---

[7]From [51], p. 43.

[8]Famous for his critique of calculus; see p. 47.

While many people found Argand's proposal interesting, it was not really used in a serious way until Gauss proposed much the same idea in 1831 and showed that it could be useful mathematically. It was also Gauss who proposed the term "complex number" (by which he meant a number that has more than one component: a real part and an imaginary part). A couple of years later, William Rowan Hamilton showed that one could *start* with the

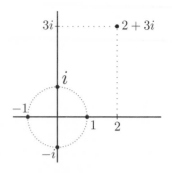

Display 2

plane, define the sum and product of ordered pairs in a convenient way, and end up with something identical to the complex numbers. Hamilton's approach completely avoided the mysterious "i"; it was just the point $(0, 1)$.

Argand and Hamilton probably saw their ideas as one more application of complex numbers: One can use them to do plane geometry. But the concreteness of the plane helped remove some of the concerns about them. That was good, because complex numbers are incredibly useful. Euler and Gauss used them to solve problems in algebra and number theory. Hamilton used complex numbers to do physics. Cauchy and Gauss devised a version of the calculus which applied to complex numbers. This "complex calculus" turned out to be extremely powerful. In the hands of Riemann, Weierstrass, and others, it became a powerful mathematical tool that played a central role in both pure and applied mathematics.

The power of these discoveries was captured well by French mathematician Jacques Hadamard, who said that "the shortest path between two truths in the real domain passes through the complex domain." Even if we care only about real number problems and real number answers, he argued, the easiest solutions often involve complex numbers.

So why should we "believe" in complex numbers? Because they are so useful!

For a Closer Look: Most of the big historical reference books include discussions of the history of complex numbers. For more detail, see chapter 3 of [69]. See [121] for an invitation to imagine what complex numbers might be and [130] for a more technical account of "the story of $\sqrt{-1}$." The article [111] offers an interesting discussion of what exactly Bombelli thought about his "radicals of a new kind."

18   Half is Better
Sine and Cosine

The story of the sine function goes back as least as far as the Greek astronomer Hipparchus of Rhodes in the 2nd century B.C. Like other Greek astronomers, he wanted to come up with a model that would describe how stars and planets move through the night sky. The sky was represented as a gigantic sphere (we still speak of the "celestial sphere"), and the positions of stars were specified by angles. Working with angles is difficult, so it turned out to be useful to relate the angle to some (straight) line segment. The segment they chose was the *chord*. As shown in Display 1, a central angle β in a circle of some fixed radius determines a chord, and we call it (or its length) the chord of β. Using chords, it was possible to compute current and future positions of stars and planets.

It is generally believed that Hipparchus constructed a table of such chords. He apparently worked with a circle of radius 3438 and then wrote out a table giving the lengths of the chords corresponding to various different angles. (Why 3438? Because then the circumference is very close to $21600 = 360 \times 60$, so that each minute of arc corresponds to approximately one unit of length on the circumference. This makes $\sin(0°\ 1') \approx 1$.) This table has not sur-

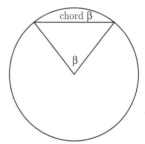

The chord of an angle

Display 1

vived, so we don't know exactly how the chords were computed. We know about it from references by other Greek mathematicians.

The greatest of the ancient Greek astronomers was surely Claudius Ptolemy. In his *Almagest*, written in the 2nd century A.D., we can learn the beginnings of the theory of chords. Most of the first chapter of this book is dedicated to proving basic theorems about chords and how they can be used to get information about "spherical triangles," triangles made by great circles on the surface of a sphere. In addition to working out theorems, Ptolemy explained how to construct a table of chords. Starting out with a few exact results, he then devised a method that allowed him to compute approximations to the chords of angles from $\frac{1}{2}°$ to $180°$. These went into his table.

The next important step was taken in India. In a work written in the early 5th century A.D., we find a table of "half-chords." This reflected an important insight. While the chord may be the simplest way to associate a line segment to an angle, it turns out that in many situations what one needs to use is *half the chord of twice the angle*. This breaks the isosceles triangle associated with the chord into two right triangles, which are easier to work with. Indian astronomers understood this early on and therefore moved from tabulating chords to tabulating half the chords. The word for "chord" was *jyā*, which means "bow-string" (can you see the bow and the bow-string in Display 2?). Half the chord should be *jyā-ardha*, but since they consistently used only the half-chord they often simply used *jyā* or the variant form *jīvā*.

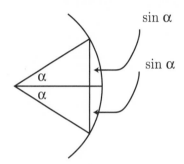

sin α

sin α

The sine is half the chord
of twice the angle.

Display 2

As you can see from Display 2, the half-chords of the Indian mathematicians are exactly the same as our sines. You can also see the other Indian trigonometric functions in the diagram. The cosine is the segment from the center to the bow-string; it was simply "the sine of the complement." The segment from the bow-string to the circle was referred to as the "arrow" (of course). We would call it the "versine", which for us is just $1 - \cos(\alpha)$. The tangent was known to Indian mathematicians, but in a different context (see Sketch 28).

There was one big difference between the Indian functions and our own, however. We think of the sine as a ratio between the segment indicated in the picture and the radius of the circle. They thought of the half-chord and the arrow as the actual line segments in a circle of a particular radius. (Ptolemy, for example, does all his computations with a circle of radius 60.) In other words, the half-chord is a segment whose length is $R \sin \alpha$. Whenever any of the trigonometric segments were used, one had to take into account the radius of the circle, scaling it appropriately if the situation required a different radius. The earliest Indian tables used a circle with radius 3438. That number suggests the influence of Hipparchus, of course, but some historians argue that a radius of $(360 \times 60)/2\pi$ is such a natural choice that it could have been made independently.

Indian mathematicians developed very sophisticated methods for computing tables of half-chords. Since there is no way to compute exactly the length of a chord of an arbitrary angle, these methods were approximation techniques. From Āryabhaṭa in the 6th century to Bhāskara in the 12th century and onward, one finds more and more sophisticated methods for approximate computations. Many of these methods anticipate ideas that were later rediscovered by European mathematicians.

In almost every case, Indian mathematical ideas reached Europe by way of the Arabic mathematicians. That's how it was with Indian trigonometry. The Arabs learned their astronomy from India, and this included learning about tables of half-chords. Rather than translate the Sanskrit word, however, the Arabic mathematicians simply invented a word, turning the Sanskrit *jīvā* into *jiba*. For the Indian "arrow," however, they just chose the Arabic word for arrow.

The choice of the nonsense word *jiba* was a problem. Because Arabic is written without vowels, the word was written simply as *jb*. It was inevitable that readers would see it as an actual word, *jaib*, which mean could mean "cove," "bay," or "pocket." In his astronomy book *The Canon*, al-Bīrūnī says *jaib* was the Arabic equivalent to the Sanskrit *jiva*, so that version was already current by the late 10th century.

As they did in every other case, the Arabic mathematicians added their own ideas to the subject. In fact, Arabic trigonometry became quite sophisticated. They discovered the connections between trigonometry and algebra. For instance, they knew that solving cubic equations would help solve the problem of computing sines of arbitrary angles. They deepened and expanded what was known about spherical triangles, making computational astronomy easier. They also added other functions to the theory (still thought of as specific segments, however). We discuss that development in Sketch 28.

When European mathematicians discovered this material, there was, as usual, a rush to translate and study the Arabic works. When it came time to translate *jaib*, the translators chose the Latin word *sinus*, which originally meant "bosom," had come to be applied to the hollow created by the fold of a tunic at the bosom, and from that to any hollow of that shape, including a cove or a bay. (Our word "sinuous" descends from the same Latin root: something is sinuous if it has lots of coves and hollows.) This is how we get the word "sine." For the "arrow," the word used at first was the Latin equivalent, *sagitta*. It's a pity that later "sagitta" became the much more boring "versine" — it would be more fun to talk about the "arrow."

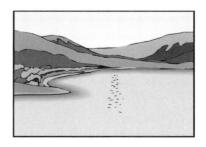

During most of this period, the main application of trigonometric ideas was in astronomy. Astronomers mostly use spherical trigonometry, so it was this topic that filled most of the books. In the 15th century, however, trigonometry began to become an object of interest in itself. The most important trigonometrical work from that period was a book by Johannes Müller, who is usually known as Regiomontanus because he was born in the city of Königsberg. ("Königsberg" means "the king's mountain," and "Regiomontanus" is Latin for "from the king's mountain.")

Regiomontanus wrote *On All Sorts of Triangles*[1] around 1463, but the book wasn't published until several decades later. Though it is clear that he knew about the Arabic work on the tangent function, in his book he uses only the sine. The book contains the basic theory and applications to the geometry of both plane and spherical triangles. For Regiomontanus, the sine is still not a ratio; as in the ancient Indian works, it is the length of a particular line segment. The book includes a large table of sines, computed with respect to a circle of radius 60,000. This radius was known as the "total sine," and the computations had to take it into account.

What about the cosine? Well, every so often one needed to use the sine of the complementary angle, that is, one needed $\sin(90° - \alpha)$. (See Display 3.) At this point, however, no one seems to have given that quantity a special name. It was just the sine of the complement. Two centuries later, however, *sinus complementi* had become *co. sinus* and then *cosinus*.

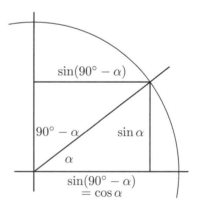

The sine of the complement

Display 3

Regiomontanus's book was enormously influential. Over the next few decades, several other books on the subject were written. Some of them were just reworkings of the material in Regiomontanus, but a few introduced new ideas. In the mid-16th century, Georg Joachim Rheticus explained how to define the sine and other functions in terms of right triangles, without reference to an actual circle. (See Sketch 26.) Thomas Fincke invented the words *tangent* and *secant*.

---

[1] The original title is in Latin: *De Triangulis Omnimodis*.

Finally, Bartholomew Pitiscus invented the word *trigonometry* and used it in the title of his book, first published in 1595. The title page of Pitiscus's book, *Trigonometry, or the Measurement of Triangles*, advertises that it will include material that can be applied to surveying, geography, and astronomy. (See Display 4.) Pitiscus's book became a standard reference and textbook, and established trigonometry as an independent mathematical topic with many different applications.

Trigonometry continued to be very popular in the 17th century. This was the time of the rise of algebra (see Sketch 8), and trigonometry offered a way to use algebraic techniques to solve geometric problems. It was also sometimes used to solve algebraic problems. For example, François Viète showed that one could solve certain cubic equations using trigonometric functions, neatly reversing what the Arabic mathematicians had done centuries before.

Title page of the 3rd edition of Pitiscus's trigonometry

Display 4

All of this trigonometry still looked very different from what we learn today. For one thing, the sine was still a particular line segment drawn in a circle of a particular radius, rather than a ratio. For another, no one had yet thought of the sine as a *function* in the modern sense. Both of these changes happened only after the calculus was invented, and they were really cemented in place by Leonhard Euler in the 18th century. It was Euler who convinced people that they should think of the sine as a function of the arc in a unit circle (that is, that they should think of it as a function of an angle measured in radians). Euler's influence was great, and it is because of his work that we approach trigonometry as we do today.

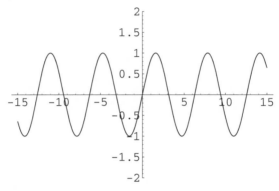

Graph of $y = \sin x$

Display 5

What about the sine curve, the graph of the sine, like the curve in Display 5? Well, in the 17th century, Gilles de Roberval sketched a sine curve when he was computing the area under a cycloid. It was only at this point that the sinuous form of the sine curve became clearly visible.

For a Closer Look: We first learned the etymology of the word "sine" from a chapter in V. Frederick Rickey's (still unpublished) "Historical Notes for the Calculus Classroom." You can see a more recent discussion in [140, Section 8.1.2]. The best book on the early history of trigonometry is [175]. Maor's [119] is a readable popular account that emphasizes the more recent history. A shorter account of the history of trigonometry is [69, ch. 1]. The copy of the book whose cover page is shown in Display 4 is in the Colby College Special Collection; the photograph was taken by Margaret Libby.

19 Strange New Worlds
The Non-Euclidean Geometries

Euclid's systematic approach to plane geometry (see Sketch 14) was so good that it took more than 2000 years to unravel a mystery right at its heart. This sketch outlines the story of that mystery and the search for its solution. It begins with Euclid's Fifth Postulate, which says:

> If a straight line falling on two straight lines makes the sum of the interior angles on the same side less than two right angles, then the two straight lines, if extended indefinitely, meet on that side on which the angle sum is less than the two right angles.

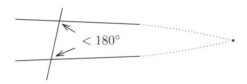

This postulate doesn't seem to be about parallel lines. It states a property that guarantees that a pair of lines is *not* parallel. But because it plays a crucial role in proving many properties of parallel lines, it is usually called the *Parallel Postulate*.

One of Euclid's initial definitions says that *parallel lines* are (straight) lines in the same plane which, if extended indefinitely in both directions, do not meet one another in either direction. But he doesn't deal with parallel lines until much later.

In fact, the postulate doesn't get used until Proposition 29. Why did Euclid organize his propositions in this way? No one really knows, but most people have assumed he knew that there was something unusual about the Fifth Postulate and so did as much as possible without it. Once he started using it, however, Euclid used it with power. The remaining twenty propositions of his Book I establish the essential properties of parallel lines, parallelograms, and squares, including the familiar fact that parallel lines are everywhere equidistant. He also used it to prove some fundamental geometric ideas whose connection to parallel lines is not so obvious, including the Pythagorean Theorem and the fact that the sum of the angles of a triangle equals exactly 180° (two right angles).

In some ways, the Parallel Postulate seems more like a theorem. It is not as simple or self-evident as his other four postulates, and yet there was no doubt that it was true. Euclid was masterful at proving theorems and organizing ideas. Why was he unable to *prove* this statement? For more than 2000 years, most mathematicians were convinced that it was because Euclid just wasn't quite clever enough. They were convinced that it was really a theorem, and many people proposed proofs. The 5th-century Greek philosopher Proclus argued that it *was* a theorem and discussed possible ways of proving it, without finding a convincing argument. After Arab scholars of the 8th and 9th centuries translated classical Greek works into Arabic, they also attempted to prove the Parallel Postulate. This quest continued in both the Middle East and the West for many centuries. But in every case the proposed proofs were flawed. A correct understanding of this maddening postulate that looked like a theorem did not emerge until the 19th century.

As they tried to prove the Parallel Postulate, some mathematicians found logically equivalent substitute statements that they considered clearer or easier to work with. The most famous of these substitutes is known as *Playfair's Postulate*, named after the Scottish scientist John Playfair, who made it popular in the 18th Century.

Playfair's Form of the Parallel Postulate: Through a point not on a line, there is exactly one line parallel to the given line.

Playfair's Postulate is so well known that many current geometry books present it as the Parallel Postulate, in place of Euclid's original statement. For the rest of this sketch, we shall do likewise.

Early in the 18th Century, an Italian teacher and scholar named Girolamo Saccheri tried a clever new approach to the Parallel Postulate question. He reasoned like this:

- Euclid's axioms do not contain any contradictions.

- We believe that the Parallel Postulate can be proved from Euclid's other axioms, but so far no one has been able to prove it.

- Suppose it can be proved. Then, replacing the Parallel Postulate by its negation puts a contradiction into the system.

- Therefore, if I use the negation of the Parallel Postulate as an axiom and find a contradiction in this new system, I will have shown that the Parallel Postulate can be proved from the other axioms, even though I don't actually have a direct proof of it!

Saccheri began with a rectangle-like figure: two equal line segments perpendicular to a horizontal base and connected by a segment at the top. This figure is now called a *Saccheri quadrilateral*. In Euclidean geometry it must be a rectangle, but Saccheri realized that proving this requires the Parallel Postulate. Once he proved that that the top two angles must be equal, Saccheri had three cases to consider:

1. Both angles are right.
2. Both angles are obtuse ($> 90°$).
3. Both angles are acute ($< 90°$).

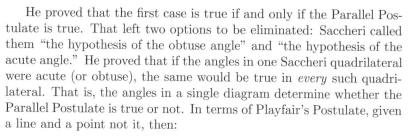

He proved that the first case is true if and only if the Parallel Postulate is true. That left two options to be eliminated: Saccheri called them "the hypothesis of the obtuse angle" and "the hypothesis of the acute angle." He proved that if the angles in one Saccheri quadrilateral were acute (or obtuse), the same would be true in *every* such quadrilateral. That is, the angles in a single diagram determine whether the Parallel Postulate is true or not. In terms of Playfair's Postulate, given a line and a point not it, then:

1. on the hypothesis of the obtuse angle, there is *never* a line through the point parallel to the line;
2. on the hypothesis of the acute angle, there is *always* more than one line through the point parallel to the line.

Thus, Saccheri had to show that each of these two cases results in a contradiction. The first one was easy. Euclid had already proved, without using the Parallel Postulate, that parallel lines must exist, yielding a direct contradiction in that case. The second case was much more stubborn. Saccheri proved many interesting results, but he could not find a clear contradiction. Finally, he considered what happened "at infinity" to get what seemed to him like a contradiction. It wasn't very convincing, and one suspects that he knew it. His results were published in 1733, in a book called *Euclides Vindicatus* (literally "Euclid Vindicated"), which was promptly forgotten for almost 100 years.

In the 19th century people started to wonder: *Can there be a plane geometry in which, through a point not on a line, there is more than one line parallel to the given line?* The great German mathematician Carl Friedrich Gauss explored this possibility around 1810, but he didn't publish any of his findings, so almost nobody knew about his work then. The first published investigation of such a different geometry appeared in 1829. It was written by Nicolai Lobachevsky, a Russian mathematician who devoted much of his life to studying it. At about the same time, János Bolyai, a young officer in the Hungarian army, was working out the same ideas, which he published in 1832. All of them came to the same surprising conclusion: If the Parallel Postulate is replaced by the assumption that there are many lines through the point that do not intersect the line, the resulting system contains no contradictions! Each had set out to prove the Postulate by assuming its negation. Instead, they discovered a different geometry with no obvious flaws.

That should have settled, once and for all, the 2000-year-old question about Euclid's fifth postulate:

> The Parallel Postulate cannot be proven from the other four
> postulates of Euclid's geometry!

Alas, very few people were paying attention. Bolyai published his results in Latin, in an appendix to a book by his father. Lobachevsky published his in Russian in an obscure journal. We know from his correspondence that Gauss was aware of both works, but very few other people seemed to be interested.

So we had an entirely new geometry — an entirely new theory of space. Of course, the space for this geometry is different from Euclidean space. Gauss named it "non-Euclidean space"; we now call it *hyperbolic space*. One of the reasons that led people to ignore it was the idea, strongly defended by some followers of philosopher Immanuel Kant, that Euclidean geometry *must* be the geometry of real space.

The most important breakthrough came in the work of Bernhard Riemann. In an 1854 lecture, he laid down an entirely new approach to geometry based on "local" information. His idea was that what we observe of space is only a very small portion. In that portion, we have a way to measure distances. To measure larger distances, Riemann used integration, adding up the small distances along a curve. He explained the basic conditions that local distances have to obey and argued that *any* local distance satisfying those conditions would yield a reasonable geometry. He showed that a local distance determines a local *curvature*. To get Euclidean geometry, one just had to make the curvature constant and equal to zero. He didn't say so in the lecture, but to get hyperbolic geometry, one just has to make the curvature constant and negative.

What about positive curvature? Well, in a space with positive curvature, there would be no parallel lines at all! In other words, a space with positive curvature would be one where Saccheri's hypothesis of the obtuse angle holds. But Saccheri had found a contradiction in this case, hadn't he? And Euclid's proof that parallel lines exist didn't use the Parallel Postulate. So some *other* postulate must fail in this case.

One of Euclid's postulates says that lines can be "extended continuously." Both Saccheri and Euclid had both assumed that lines were infinitely long. Riemann observed that "extended continuously" did not necessarily imply that. For instance, an arc of a circle can be extended continuously as far as we please; it has no endpoint, but its length is finite. (Of course, extending the arc may mean retracing a part of it.)

That's all Riemann said, but others quickly saw that replacing Euclid's assumption of infinite lines with the weaker assumption that lines had no boundary points led to a new geometry, now called *elliptic geometry*, which satisfies Euclid's first four postulates but not the Parallel Postulate.

To visualize a model of elliptic geometry, use the surface of a sphere as the "plane" and locations on that sphere as "points." The "lines" of this geometry are the *great circles*, circles that divide the sphere into two equal parts, like the equator or the longitude lines on the Earth. Such circles are called "great" because they are the largest circles that can be drawn on the sphere. This means that the shortest path between any two points on the sphere is an arc of the great circle through those points, so these circles are analogous to the straight lines of the Euclidean plane. Now, any two great circles must intersect (think about trying to cut the Earth exactly in half without crossing the Equator), so this geometry has no parallel lines.[1]

Thus, by the middle of the 19th century there were three different "brands" of geometry, distinguished from one another by the way they treated parallel lines. In the 1870s, Felix Klein emphasized that one should consider all three options together. Such geometries could have (constant) positive, negative, or zero curvature. He called the three cases "elliptic," "hyperbolic," and "parabolic," respectively,[2] but only the first two names stuck. If the curvature was zero, the geometry was Euclidean. He called both other cases *non-Euclidean geometries.*

All three of these geometries are consistent systems, but their disagreement about parallelism gives them strikingly different properties. For instance, only in Euclidean geometry are there triangles that are similar but not congruent. In the non-Euclidean geometries, if the corresponding angles of two triangles are equal, then the triangles *must* be congruent. Another peculiar difference among these geometries is that the angle sum of a triangle varies according to the type of geometry:

in Euclidean geometry, it is always exactly 180°;

in hyperbolic geometry, it is always less than 180°; and

in elliptic geometry, it is always greater than 180°.

In the latter two cases, the amount of deviation from 180° varies with the area of the triangle.

---

[1] This simplified description ignores a technical difficulty: Diametrically opposite points must be treated as identical, otherwise two points do not determine a unique line (great circle). An appropriate equivalence relation disposes of this issue.

[2] These terms refer to differences in curvature, not to conic sections.

Looking at this picture, you might think that the second and third figures are not really triangles because their sides don't look straight. But they are. They are the shortest paths between the vertices *in the space of that geometry*. They appear curved because "flattening out" the triangles to represent all three of them on this page (which is a Euclidean plane) distorts the relative distances between points on the non-Euclidean triangles. To get a better feeling for this, visualize the third triangle as being drawn on a balloon. On that surface it becomes clearer that the sides are the shortest paths between the vertices.

One more difference is worth noting because it seems unrelated to parallel lines, or even to straight lines. The ratio of the circumference, C, of a circle to its diameter, d, depends on the type of geometry, too:

> in Euclidean geometry, it is exactly π;
> in hyperbolic geometry, it is always greater than π;
> in elliptic geometry, it is always less than π.

Again, the amount of deviation varies with the area.

Faced with three conflicting geometric systems, it is tempting to pick the old, familiar one as the "true" geometry and think of the other two as made-up oddities. But is that right? In fact, the question is unanswerable. The real issue is not "what's true," but "what works." Geometries are tools, designed by humans, to help us deal with our world. Like any other tools, some are suitable for one job, some for another. If you are a builder, surveyor, or carpenter, then Euclidean geometry is by far the simplest to use; it works well for such things. If you are an astronomer studying distant galaxies or a theoretical physicist, you will need to take account the curvature of space, and so will need a different geometry. A geometry is a tool to be chosen by the worker, not a fixed feature of the job site.

For a Closer Look: Lots of books discuss the history of non-Euclidean geometry. [137] is a well written account of how geometric ideas affect our understanding of the universe; chapters III to V deal with non-Euclidean geometry. Also useful are lectures 26 and 27 in [54]. An interesting article on attempting to prove the Parallel Postulate is [71]. For a deeper account of non-Euclidean geometries, see [78] and [80].

In the Eye of the Beholder
Projective Geometry

20

As the liberating trends of the Renaissance spread throughout Europe, prompting scientists and philosophers to explore the world around them with renewed vigor, artists searched for ways to mirror that reality on paper and canvas. Their main problem was perspective — how to portray depth on a flat surface. The artists of the 15th century realized that their problem was geometric, so they began to study the mathematical properties of spatial figures as the eye sees them. Filippo Brunelleschi made the first intensive efforts in this direction, and soon other Italian painters followed suit.

The most influential Italian artist in the study of mathematical perspective was Leone Battista Alberti, who wrote two books on the subject. It was he who proposed the principle of painting what one eye sees. That is, he thought of the surface of a picture as a window or screen through which the artist views the object to be painted. As the lines of vision converge to the point where the eye is viewing the scene, the picture on the screen captures a cross section of them. (See Display 1.)

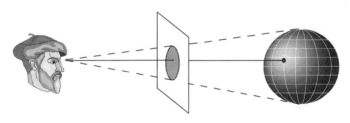

Display 1

Alberti developed a number of mathematical rules for applying this principle. He also posed a fundamental question: If an object is viewed from two different locations, then the two "screen images" of that same object will be different. How are those images related, and can we describe their relationship mathematically? The screen images are called *projections* of the object; the study of how projections are related became the motivating question for a new field of mathematics called *projective geometry*. Among the most prominent people of the 15th and early 16th centuries who studied, used, and advanced this mathematical theory of perspective were Italian artists Piero della Francesca and Leonardo da Vinci, and also the German artist Albrecht Dürer, who wrote a widely used book on the subject.

Now think of projections "in the other direction," so to speak. That is, think of the "eye" as a light source that projects a figure on one plane to its image on another, just like a slide projector shows a figure on a film slide as an image on a screen. If you tip the screen, you can distort the image in lots of different ways. (See Display 2.) You can change distances and angles, for instance, but there are some fundamental properties of figures that you can't change. For example, a straight line will always project to a straight line.

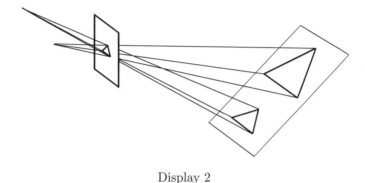

Display 2

Here's a more striking example: The image of a circle may not be a circle, but it will *always* be a conic section.[1] In fact, the image of *any* conic section will always be a conic section! This remarkable property of conic sections was the basis for a groundbreaking study of projections by French engineer and architect Girard Desargues. Desargues's work was generally overlooked in his day but was rediscovered and finally appreciated in the mid-19th century. By then, Jean Victor Poncelet had published a very influential book on projective geometry, which he had planned (without benefit of any reference books) while he was a Russian prisoner of war after Napoleon's defeat at Moscow.

Building on the work of Desargues and Poncelet, a number of 19th-century French and German mathematicians turned projective geometry into a major field of study. Their mathematical generalizations of the ideas of perspective and projection led to some surprising insights. One of the most striking and powerful is the *principle of duality*.

---

[1] The conic sections are circles, ellipses, parabolas, and hyperbolas. See Sketch 28.

To explain this idea, let's step back for a moment to the artist's view of projections. A well-known example of perspective drawing is the image of railroad tracks stretching away to a distant horizon. The tracks, which are parallel, appear to meet at a point at or just beyond the horizon. In fact, on the plane of the artist's canvas, these lines *do* meet. That is, in order for these lines to appear to be the same distance from each other

as they recede from the viewer, they must meet "at infinity." All the lines that appear to be parallel to a particular line should meet at the same point at infinity. Projective geometry takes those points at infinity seriously. The plane of two-dimensional projective geometry is the regular Euclidean plane with an extra line added — an *ideal line* that contains exactly one point for each "family" of parallel lines in the Euclidean plane.[2] In this way, *every* pair of straight lines in the projective plane crosses at exactly one point.

This brings us back to duality. In Euclidean geometry, "Two points determine exactly one line" is a well-known fact. It is true in projective geometry, too, and so is "Two lines determine exactly one point." In fact, any statement about points and lines that is true in projective geometry *remains true if the words "point" and "line" are interchanged* (assuming that you make the appropriate adjustments in terminology). This is the *principle of duality*, and each of the two statements is called the *dual* of the other. It was first recognized in the middle of the 19th century and established definitively early in the 20th century by constructing an axiom system for projective geometry in which the dual of each axiom is also an axiom.

For example, here is an axiom system for plane projective geometry in which each axiom is its own dual:[3]

1. Through every pair of distinct points there is exactly one line, and every pair of distinct lines intersects in exactly one point.

2. There exist two points and two lines such that each of the points is on only one of the lines and each of the lines is on only one of the points.

3. There exist two lines and two points not on those lines such that the intersection point of the two lines is on the line through the two points.

---

[2]Each family of parallel lines can be related to a single real number, in much the same way as all the lines of each family are related to their common slope.

[3]See [58], p. 116.

The fact that Axiom 3 is its own dual may not be obvious at first. However, keep in mind that the intersection point of two lines is the dual of the line through two points. Now if you interchange "point" and "line" in Axiom 3 and make that language adjustment, you can see that the resulting statement says the same thing: "There exist two points and two lines not on those points such that the line through the two points passes through the intersection point of the two lines."

The dual statements we have presented so far don't illustrate the power of duality very well; they're too simple. Here is a more interesting example, the so-called "mystic hexagram" theorem of Blaise Pascal, which stemmed from the work of Desargues on conic sections:

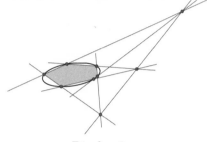

A hexagon can be inscribed in a conic section if and only if the points (of intersection) determined by its three pairs of opposite sides lie on the same (straight) line. (Display 3 illustrates Pascal's theorem for an ellipse.)

Display 3

Applying point-line duality, the sides of a polygon become the vertices and vice versa, and "inscribed" (vertices lie on the conic) becomes "circumscribed" (sides are tangent to the conic). Thus, the dual of Pascal's theorem is:

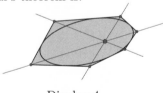

Display 4

A hexagon can be circumscribed about a conic section if and only if the lines determined by its three pairs of opposite vertices meet at the same point. (Display 4 illustrates the dual of Pascal's theorem for an ellipse.)

A proof of this theorem was first published in 1806, more than 175 years after Pascal (at the age of 16) had stated his mystic hexagram theorem but before the principle of duality was understood. By duality, we now know that any proof of either theorem *automatically* guarantees the truth of the other one, too!

For a Closer Look: See [58] for more on projective geometry. Look at [61] and [6] for the early history of perspective. For the history of projective geometry in the 19th century, the best source is [78].

21 What's in a Game?
The Start of Probability Theory

In 1654 the Chevalier de Méré, a wealthy French nobleman with a taste for gambling, proposed a gaming problem to the mathematician Blaise Pascal. The problem was how to distribute the stakes in an unfinished game of chance. The "stakes" are the amounts of money each gambler bets at the start of a game. By custom, as soon as these bets are made, that staked money belongs to no one until the game is over, at which time the winner or winners get it all. De Méré's question, now known as the "Problem of Points," was how to divide the stakes of an unfinished game if the partial scores of the players are known. In order to be "fair," the answer should somehow reflect each player's likelihood of winning the game if it were to be finished. Here's a simple version of De Méré's Problem of Points.[1]

Xavier and Yvon have staked $10 each on a coin-tossing game. Each player tosses the coin in turn. If it lands heads up, the player tossing the coin gets a point; if not, the other player gets a point. The first player to get three points wins the $20. Now suppose the game has to be called off when Xavier has 2 points, Yvon has 1 point, and Xavier is about to toss the coin. What is a fair way to divide the $20?

The actual Problem of Points considered by Pascal asks that question for all the possible scores in an interrupted game of this kind. Pascal communicated the problem to Pierre de Fermat, another prominent French mathematician, and from their correspondence a new field of mathematics emerged. Using somewhat different methods, these two mathematicians arrived at the same answer to the problem. Here is Pascal's way of answering our simple case:[2]

A fair coin is equally likely to turn up heads or tails. Thus, if each player had two points, each would be equally likely to win the game on the next toss, so it would be fair for each player to get $10, half of the staked amount at that stage. In this case Xavier has 2 points and Yvon 1. If Xavier tosses the coin and wins, he has 3 points and hence gets the $20.

---

[1] Adapted from [1], p. 14.

[2] [1], p. 243

If Xavier loses, then each player has 2 points and hence each is entitled to $10. So Xavier is guaranteed at least $10 at this stage. Since it is equally likely that Xavier would win or lose on this toss, the other $10 should be split equally between the players. Therefore, Xavier should get $15 and Yvon $5.

Pascal handled the other cases of the interrupted game in turn, reducing each one to a previously solved situation and dividing the money accordingly. Once Pascal and Fermat found that both of their methods led to the same answers, their correspondence petered out. Unfortunately, their work did not become known until much later. But the question was in the air, and soon other scholars took up the challenge of analyzing gambling games.

Placing a numerical measure on the likelihood that something unknown might happen or might have happened is the central idea of probability. The key to understanding this process begins with the idea of equally likely outcomes, as Pascal's solution suggests. If a situation can be described in terms of possible outcomes that are equally likely, then the probability that one of them might happen is just 1 divided by their total number. This principle was recognized and explored by Girolamo Cardano more than a century before the Chevalier de Méré was playing dice, but his book on the subject, *Liber de Ludo Aleae* (Handbook on Games of Chance), was not published until nine years after Pascal and Fermat had solved the Chevalier's problem. In that book, Cardano recognized a related principle that we now call the Law of Large Numbers. In terms of equally likely outcomes, this principle is simply an affirmation of our common sense:

> If a game (or other experiment) with n equally likely outcomes is repeated a large number of times, then the actual number of times each outcome actually occurs will tend to be close to $\frac{1}{n}$. The more times the game is played, the closer the results will come to matching this ratio.

 If a single fair die is thrown, any one of its six faces is equally likely to turn up. Thus, for example, we have one chance in six of throwing a 5; its probability is $\frac{1}{6}$. This doesn't guarantee that a 5 will turn up exactly once if we throw the die six times, but the Law of Large Numbers says that, if we throw the die 100 or 1000 or 1,000,000 times, the number of times that a 5 appears will tend to get closer and closer to $\frac{1}{6}$ of the total number of throws.

Assigning probabilities to the outcomes of such situations depends on being able to count accurately the total number of *equally likely* possibilities. This can be a bit tricky sometimes. For instance, there are eleven possible outcomes for throwing a pair of dice (getting 2 or 3 or... or 12 spots), but they're not all equally likely. If you count up all the possible pairings of the numbers 1 through 6 for each die, six of them add up to 7, but only one of them adds up to 12. Thus, the probability of throwing 7 is $\frac{6}{36}$, or $\frac{1}{6}$, but the probability of throwing 12 is only $\frac{1}{36}$. In some sense, each of the different numerical occurrences must be "weighted" differently by accounting for the number of different equally likely ways it might occur.

The principle of weighting outcomes in this way can be extended to situations in which counting equally likely possibilities won't work. For instance, a spinner at the center of a disk with red, yellow, and blue segments of different sizes is not equally likely to stop on  any particular color. Rather, the size of each colored segment should "weight" the likelihood of its occurrence; if half the disk is colored red, then the probability of a fair spinner stopping on red should be $\frac{1}{2}$, and so on. The underlying principle of probability, recognized by Cardano, Pascal, and Fermat, is that the probability measurement assigned to each possible outcome should be a number less than 1 (reflecting its relative likelihood of occurring) and that the numbers for all the possible outcomes in a situation should add up to exactly 1. (If you think of probabilities as "likelihood percentages," then the total of all of them should be 100%.)

In 1657, the Dutch scientist Christiaan Huygens became aware of the ideas of Pascal and Fermat and began to work more systematically on the question. The result was *On Reasoning in a Dice Game*, which extended the theory to games involving more than two players. Huygens's approach started from the idea of "equally likely" outcomes. His central tool was not the modern notion of probability, but rather the idea of *expectation* or "expected outcome." Here is a simple example.

> You are offered one chance to throw a single die. If 6 comes up, you get $10; if 3 comes up, you get $5; otherwise, you get nothing. What is a fair price to pay for playing this game?

From the modern point of view, the mathematical expectation of a game is found by adding the products of each possible reward multiplied by the probability that you will get it. In this case, each of the six faces is equally likely to turn up (assuming the die is fair), so you have one chance in six of getting \$10, one chance in six of getting \$5, and four chances in six of getting nothing. Therefore, the mathematical expectation is

$$\tfrac{1}{6} \cdot \$10 + \tfrac{1}{6} \cdot \$5 + \tfrac{4}{6} \cdot \$0 = \$2.50$$

This means that, if a casino were to offer this game to its customers for a fee of \$2.50, it would expect to break even in the long run. If it charged \$3 to play the game, it should expect to make \$.50 per player, in the long run. (If you buy a \$1 lottery ticket and use the data on its back to calculate your mathematical expectation, you'll find that it's considerably less than \$1. That's why states run lotteries.) Huygens reversed this process, using the expectation to compute the probability instead of the other way around. But the fundamental idea was the same: Equally likely outcomes mean equal expectations.

Mathematical expectation, like most of probability theory, applies to far more than lotteries and casino gambling. Among other things, it is fundamental to the way insurance companies assess their risks when they underwrite policies. Jakob Bernoulli recognized the wide-ranging applicability of probability in his book *Ars Conjectandi* ("The Art of Conjecture"), which was published in 1713, eight years after his death. There are many things in the book. In the fourth chapter, Bernoulli examined the relationship between theoretical probability and its relevance to various practical situations. In particular, he recognized that the assumption that outcomes were equally likely was a serious limitation when discussing human life spans, health, and the like, suggesting instead an approach based on statistical data. In so doing, Bernoulli also sharpened Cardano's idea of the Law of Large Numbers. He asserted that if a repeatable experiment had a theoretical probability p of turning out in a certain "favorable" way, then for any specified margin of error the ratio of favorable to total outcomes of some (large) number of repeated trials of that experiment would be within that margin of error. By this principle, observational data can be used to estimate the probability of events in real-world situations.

If we want to do this with some precision, we need to be able to decide how many observations are needed. Bernoulli attempted to do this in his book but ran into serious problems. The mathematics was just very hard! He did manage to estimate a number of trials that he could show was enough, but the number he got was huge — so big that it must have been a great disappointment. If getting a reasonable estimate of a probability requires a ridiculously large number of trials, then it can't really be done in practice. Perhaps this is why Bernoulli's book remained unpublished until after he died. Once it was published, however, other mathematicians managed to improve on his methods and show that the number of observations didn't need to be as large as Bernoulli had thought.

The probabilistic point of view was not easily accepted. Consider life insurance, for example. We think it's obvious that probabilistic thinking will help companies make money selling life insurance. In particular, because we believe in the Law of Large Numbers, we understand that a company is better off if it sells many policies. The more policies it sells, the more likely it is that the death rates will be as expected, so that the company will make a profit. In the 18th century, however, many companies seemed to feel that each new policy sold increased the risk to the company. Hence, they felt that selling too many policies was positively dangerous!

Interest in probability questions led to a variety of results by a variety of people during the 18th century. Towards the end of that century, Pierre Simon Laplace, a French mathematician of wide-ranging interests and prodigious talent, became interested in probability questions. He wrote a series of papers on the subject between 1774 and 1786, before focusing his efforts on the mathematics underlying the workings of the solar system. In 1809, Laplace returned to probability by way of a statistical question, the analysis of probable error in scientific data gathering. Three years later, he published *Thèorie Analytique des Probabilités* ("Analytical Theory of Probabilities"), an encyclopedic *tour de force* that pulled together everything he and others had done in probability and statistics up to that point. It was truly a masterwork, but its technical, dense style made much of it inaccessible to all but the most determined, mathematically sophisticated reader. British mathematician Augustus De Morgan wrote of it:[3]

The *Thèorie Analytique des Probabilités* is the Mont Blanc of mathematical analysis; but the mountain has this advantage over the book, that there are guides always ready near the former, whereas the student has been left to his own method of encountering the latter.

---

[3]From an 1837 review, as quoted in [20], p. 455.

To make his ideas more accessible to a wider audience, Laplace wrote an expository 153-page preface to the second edition in 1814. This preface, which contained very few mathematical symbols or formulas, was also published as a separate booklet, entitled *Philosophical Essay on Probabilities*. In it, Laplace argued for the applicability of mathematical probability to a wide range of human activities, including politics and what we now think of as the social sciences.

In this respect he was echoing the ideas of Jakob Bernoulli, whose *Ars Conjectandi* of a century earlier had suggested ways of applying probabilistic principles to government, law, economics, and morality. As the study of statistics has developed in the past two centuries, it  has provided the means by which the vision of Bernoulli and Laplace has become a reality. Today the ideas of probability are applied not only to the fields they suggested, but to education, business, medicine, and many other areas. (For information about the history of statistics, see Sketch 22.)

For a Closer Look: See [44] for an account of the Fermat-Pascal correspondence and its outcome. There are several scholarly accounts of the history of probability: [168] is more mathematical, while [37] and [82] take (each in a different way) a broader view.

22 Making Sense of Data

Statistics Becomes a Science

S*tatistics* is a word used in a wide variety of senses and often invoked to lend credibility to otherwise doubtful opinions. We sometimes use it to refer to *data*, especially numerical data — things like "52% of Americans like blue M & M's" or "93% of statistics are made up." When used in this sense, *statistics* is plural: each bit of data is a "statistic."[1] When *statistics* is singular, it refers to the science that produces and analyzes such data. This science has deep historical roots, but it really blossomed in the early 20th century.

The gathering of numerical data — herd sizes, grain supplies, army strength, etc. — is a truly ancient tradition. Tabulations of this sort can be found among the earliest surviving records of early civilizations. They were used by political and military leaders to predict and prepare for possible famines, wars, political alliances, or other affairs of state. In fact, the word *statistics* comes from *state*: it was coined in the 18th century to mean the scientific study of the state and quickly shifted to focus on political and demographic data of interest to the government.

Such data-gathering has existed ever since there have been governments (in fact, some scholars see the need for such data as one reason for the invention of numbers themselves). But only in the past few centuries have people begun to think about how to analyze and understand data. We begin the story in London in 1662, when a shopkeeper named John Graunt published a pamphlet entitled *Natural and Political Observations Made upon the Bills of Mortality*. The Bills of Mortality were weekly and yearly burial records for London, statistics (plural) which had been gathered by the government and filed away since the mid-16th century. Graunt summarized these records for the years 1604–1661 as numerical tables. He then made some observations about the patterns he observed: more males than females are born, women live longer than men, the annual death rate (except for epidemic years) is fairly constant, etc. He also estimated the number of deaths, decade by decade, in a "typical" group of 100 Londoners born at the same time. His tabulated results, called the London Life Table, signaled the beginning of data-based estimation of life expectancy.[2]

---

[1] There is a technical sense of *statistic* which is more precise than this, but we're focusing on popular usage here.

[2] See [20], Ch. 9, or [88] for more about Graunt and the Bills of Mortality.

Graunt, together with his friend William Petty, founded the field of "Political Arithmetic," that is, the attempt to obtain information about a country's population by analyzing data such as the Bills of Mortality. Their approach was very unsophisticated. In particular, Graunt had no way of telling whether certain features of his data were significant or were accidental variations. The issues raised by Graunt's analysis of mortality data soon led others to apply better mathematical methods to the problem. For example, the English astronomer Edmund Halley (for whom the famous comet is named) compiled an important set of mortality tables in 1693 as a basis for his study of annuities. He thus became the founder of *actuarial science*, the mathematical study of life expectancies and other demographic trends. This quickly became the scientific basis of the insurance industry, which relies on actuaries (armed today with far more sophisticated analytical tools) to analyze the risk involved in various kinds of insurance policies.

 In the first part of the 18th century, statistics and probability developed together as two closely related fields of the mathematics of uncertainty. In fact, they are devoted to investigating opposite sides of the same fundamental situation. Probability explores what can be said about an unknown sample of a known collection. For instance, knowing all possible numerical combinations on a pair of dice, what is the likelihood of getting 7 on the next throw? Statistics explores what can be said about an unknown collection by investigating a small sample. For instance, knowing the life span of 100 Londoners in the early 16th century, can we extrapolate to estimate how long Londoners (or Europeans, or people in general) are likely to live?

The first comprehensive book on statistics and probability was Jakob Bernoulli's *Ars Conjectandi*, published in 1713. The first three of this book's four parts examined permutations, combinations, and the probability theory of popular gambling games. In the fourth part, Bernoulli proposed that these mathematical ideas have much more serious and valuable applicability in such areas as politics, economics, and morality. This raised a fundamental mathematical question: How much data is needed before one can be reasonably sure that the conclusions from the data are correct? (For example, how many people must be polled in order to predict correctly the outcome of an election?) Bernoulli showed that, the larger the sample, the more likely it was that the conclusions were correct. The precise statement he proved is now known as the "Law of Large Numbers." (He called it the "Golden Theorem.")

The reliability of data was an important issue for both the science and the commerce of 18th century Europe. Astronomy was seen as the key to determining longitude, and the reliable measurement of longitude was seen as the key to safe seagoing navigation.[3] Astronomers made large numbers of observations to determine orbits of planets. But observations are prone to error, and it became important to know how to extract correct conclusions from "messy" data. A similar problem occurred in the controversy about the shape of the Earth — whether it was slightly flattened at the poles (as Newton had asserted) or at the Equator (as the director of the Royal Observatory in Paris claimed). Resolution of this issue depended on very accurate measurements "in the field," and different expeditions often got different answers. Meanwhile, insurance companies began to collect data of all kinds, but such data included variations due to chance, and one had to somehow distinguish between what was really going on and the fluctuations caused by errors and chance variation.

In 1733, Abraham De Moivre, a Frenchman living in London, described what we now call the *normal curve* as an approximation to binomial distributions. He used this idea (later rediscovered by Gauss and Laplace) to improve on Bernoulli's estimate

A normal curve

of the number of observations required for accurate conclusions. Nevertheless, the results of De Moivre and his contemporaries were not always powerful enough to provide satisfying answers to the fundamental question in real-world situations: How reliably can I infer that certain characteristics of my observed data accurately reflect the population or phenomenon I am studying? More powerful tools for this and other applications were still more than a century away.

What was needed first was more mathematics. Specifically, probability theory had to be developed to the point where it could be applied fruitfully to practical questions. This was done by many hands throughout the 18th century. The process culminated with the publication, in 1812, of Pierre Simon Laplace's *Analytical Theory of Probabilities*, a monster of a book that collected and extended everything known so far. It was heavily mathematical, so Laplace also wrote a *Philosophical Essay on Probabilities*, which attempted to explain the ideas in less sophisticated terms and to argue for their wide applicability.

France in the early 19th century had more than one brilliant mathematician. The work of Adrien-Marie Legendre rivaled that of Laplace in scope, depth, and insight. He made important contributions to analysis, number theory, geometry, and astronomy and was a member of

---

[3]See Dava Sobel's fascinating book [166] for more about this.

the 1795 French commission that measured the meridian arc defining the basic unit of length for the metric system. To statistics, Legendre contributed a method that set the course of statistical theory in the 19th century and became a standard tool for statisticians from then on. In an appendix to a small book on determining the orbits of comets, published in 1805, Legendre presented what he called "la méthode de moindres quarrés" (the method of least squares) for extracting reliable information from measurement data, saying:

> By this method, a kind of equilibrium is established among the errors which, since it prevents the extremes from dominating, is appropriate for revealing the state of the system which most nearly approaches the truth.[4]

Soon after, Gauss and Laplace independently used probability theory to justify Legendre's method. They also recast it and made it easier to use. As the 19th century progressed, this powerful tool spread throughout the European scientific community as an effective way of dealing with large data-based studies, particularly in astronomy and geodesy.

Statistical methods began to make inroads into the social sciences with the pioneering work of Lambert Quetelet of Belgium. In 1835, Quetelet published a book on what he called "social physics," in which he attempted to apply the laws of probability to the study of human characteristics. His novel concept of "the average man," a data-based  statistical construct of the human attributes being studied in a given situation, became an enticing focal point for further investigations. But it also drew criticism for overextending mathematical methods into areas of human behavior (such as morality) that most people considered unquantifiable. In fact, with the exception of psychology, most areas of social science were quite resistant to the inroads of statistical methods in most of the 19th century.

Perhaps because they could control the experimental sources of their data, psychologists embraced statistical analysis. They first used it to study a puzzling phenomenon from astronomy: The error patterns of observations by different astronomers seemed to differ from person to person. The need to understand and account for these patterns motivated early experimental studies, and the methodology developed for them soon spread to other questions. By the late 19th century, statistics was a widely accepted tool for psychological researchers.

---

[4]Quoted by [168], p. 13.

With the many advances made in the 19th century, statistics began to emerge from the shadow of probability to become a mathematical discipline in its own right. It truly came of age in the study of heredity begun in the 1860s by Sir Francis Galton, a first cousin of Charles Darwin. Galton was part of the Eugenics movement of the time, which hoped to improve the human race by selective breeding. Hence, he was very interested in figuring out how certain traits were distributed in the population and how (or whether) they were inherited. To compensate for the inability to control the countless factors affecting hereditary data, Galton developed two innovative concepts: regression and correlation. In the 1890s, Galton's insights were refined and extended by Francis Edgeworth, an Irish mathematician, and by Karl Pearson and his student G. Udny Yule at University College, London. Yule finally molded Galton's and Pearson's ideas into an effective methodology of regression analysis, using a subtle variant of Legendre's method of least squares. This paved the way for widespread use of statistics throughout the biological and social sciences in the 20th century.

As statistical theory matured, its applicability became more and more apparent. Many large companies in the 20th century hired statisticians. Insurance companies employed actuaries to calculate the risks of life expectancy and other individually unpredictable matters. Others hired statisticians to monitor quality control. Increasingly, theoretical advances have been the work of people in non-academic settings. William S. Gosset, a statistician at the Irish brewery Guinness, was such a person early in the 20th century. Because of a company policy forbidding employees to publish, Gosset had to sign his theoretical papers with the pseudonym "Student." His most significant papers dealt with sampling methods, particularly ways to derive reliable information from small samples.

The most important statistician of the early 20th century was R. A. Fisher. With both theoretical and practical insight, Fisher transformed statistics into a powerful scientific tool based on solid mathematical principles. In 1925, Fisher published *Statistical Methods for Research Workers*, a landmark book for many generations of scientists. Ten years later, he wrote *The Design of Experiments*, a book emphasizing that, to obtain good data, one had to start with an experiment designed to supply such data. Fisher had a knack for choosing just the right example to explain his ideas. His book on experiments illustrates the need to think about how experiments are designed with an actual event: At an afternoon tea party, one of the ladies claimed that tea tasted different according to whether one poured the tea first and then added milk or did the reverse. Most of the men present found this ridiculous, but Fisher immediately decided to test her assertion. How would one design an experiment to demonstrate conclusively whether or not the lady could indeed taste the difference?

It might seem like a frivolous question, but it is quite similar to the kinds of questions that scientists and social scientists need to resolve by their experiments.[5] Medical research also depends on carefully designed experiments of this kind. Fisher's work firmly established statistical tools as a necessary part of any scientist's toolkit.

The 20th century saw the application of statistical techniques to a widening array of human affairs. Public opinion polls in politics, quality control methods in manufacturing, standardized tests in education, and the like have become commonplace features of everyday life. Computers now allow statisticians to work with truly massive amounts of data, and this is beginning to affect statistical theory and practice. Some of the more important new ideas came from John Tukey, a brilliant man who contributed significantly to pure mathematics, applied mathematics, the science of computation, and statistics.[6] Tukey invented what he called "Exploratory Data Analysis," a collection of methods for dealing with data that is becoming more and more important as statisticians have to deal with today's large data sets.

Today, statistics is no longer considered a branch of mathematics, even though its foundations are still strongly mathematical. In his history of the subject, Stephen Stigler says:

> Modern statistics... is a logic and methodology for the measurement of uncertainty and for an examination of the consequences of that uncertainty in the planning and interpretation of experimentation and observation.[7]

Thus, in only a few centuries, the seeds planted by mathematical questions about data have blossomed into an independent discipline with its own goals and standards, one whose importance to both science and society continues to grow.

For a Closer Look: The best scholarly study of the history of statistics is [168]. See [88] for short biographies of important statisticians and [151] for accessible popular history of statistics in the 20th century.

---

[5] Rumor has it that the experiment was made and that the lady did correctly identify each cup of tea. See chapter 1 of [151].

[6] Tukey was also a genius at coining new words, including "software" and "bit."

[7] [168], p. 1.

Machines that Think?
Electronic Computers

23

It's hard to believe that electronic computers didn't exist until almost the middle of the 20th century. Today they seem to be everywhere, often tucked away in tiny places, manipulating large amounts of data at the speed of light and affecting almost every aspect of our lives. But in their early days, they were large, slow, clumsy machines, well deserving of their modern collective nickname, "dinosaur." The historical foundation for these machines and all their wondrous offspring began several centuries ago, with early attempts to streamline calculation by using some kind of mechanical device.

Some would say that the story begins 5000 years ago with the abacus, a calculating device of beads and rods that is still used today. However, it may be more appropriate to trace the modern computer's family tree only as far back as 17th century Europe. In 1617, Scottish scientist John Napier designed a set of movable sticks numbered in such a way that, by sliding them in relation to each other, multiplication was done automatically. These sticks were often made of ivory and, not surprisingly, came to be known as "Napier's Bones." Not long after that (in 1630), English clergyman William Oughtred improved on Napier's scheme by inventing the slide rule, a calculating device that became the constant companion of virtually every engineer (and many others) until the middle of the 20th century.

By the early 17th century, the Hindu-Arabic base-ten system had finally replaced Roman numerals as the system of choice for writing numbers in Europe, and algorithms for doing elementary arithmetic in this system were fairly well developed. (See Sketch

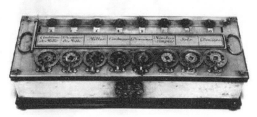

The Pascaline
(Photo courtesy of IBM Archives)

1.) In the decade between 1642 and 1652, Blaise Pascal, a brilliant French mathematician who was very young at the time, designed and eventually completed a machine for adding and subtracting which was named the *Pascaline.* Much like a car's odometer, this machine used the base-ten principle of dials numbered 0 through 9 and geared so that one full revolution on one dial would automatically move the next dial to the next number. Numbers to be added or subtracted were simply dialed in and the machine did the rest.

A major difficulty in making mechanical devices such as the Pascaline and any other precision machinery (such as reliable clocks) in the 17th century was that each gear, pivot, or other precision part had to be fashioned individually by hand. Such machines were not produced in quantity; they were crafted one by one. Thus, the genius of the inventor was often held hostage to the skill of the metalworker. So it was when Gottfried Leibniz, one of the inventors of calculus, devised an improvement on the Pascaline, a machine that could also multiply (by repeated addition) and divide (by repeated subtraction). Leibniz's machine, the *Stepped Reckoner* (shown on p. 183), represented a major theoretical advance over the Pascaline in that its computations were done in binary (base-two) arithmetic, the basis of all modern computer architecture. Nevertheless, the craftmanship of 1694 was not up to the task of making reliable copies of this machine efficiently. Its commercial success was delayed by more than 150 years. A simplified, improved version of the Stepped Reckoner developed by Charles de Colmar of France and called the *arithmometer* won a gold medal at the 1862 International Exhibition in London. Thanks to the mass-production techniques of the Industrial Revolution, it was manufactured and sold in quantity well into the 20th century. Promoters of the arithmometer claimed that it could multiply two eight-digit numbers in 18 seconds, divide a sixteen-digit number by an eight-digit number in 24 seconds, and find the square root of a sixteen-digit number in one minute. That's pretty slow by today's standards, but compared to hand calculation it was a marvel of speed and efficiency!

Early in the 19th century, Cambridge mathematics professor Charles Babbage began work on a machine for generating accurate logarithmic and astronomical tables. Besides being important for mathematics and science, these tables were essential for navigation, so the British government took an active interest in Babbage's work. By 1822, Babbage was literally cranking out tables with six-place accuracy on a small machine he called a *Difference Engine*. With government backing, he started work on a larger machine that would, he hoped, provide tables with twenty-place accuracy. After many setbacks, this project was put aside in favor of an even more ambitious plan. In 1801, Joseph-Marie Jacquard had designed a loom that wove complex patterns guided by a series of cards with holes punched in them. Its success in turning out "pre-programmed" patterns led Babbage to try to make a calculating machine that would accept instructions

Babbage's Difference Engine
(Photo courtesy of IBM Archives)

and data from punched cards. He called this proposed device an *Analytical Engine*. Like the new locomotives, which were just then harnessing steam for the first railroads, Babbage's Analytical Engine was to be steam-powered.

Babbage's assistant in this ambitious undertaking was Augusta Ada Lovelace, daughter of the poet Lord Byron and Anna Milbanke and a student of Augustus De Morgan. Lovelace translated, clarified, and extended a French description of Babbage's project, adding a large amount of original commentary. She expanded on the idea of "programming" the machine with punched-card instructions and wrote what is considered to be the first significant computer program, anticipating several modern programming devices including a "loop" for automatically repeated steps.[1] Despite the theoretical accomplishments of both Babbage and Lovelace, the Analytical Engine was never built. The metalworking technology of the time simply was not up to the precision demands of Babbage's mechanisms. Many of Babbage's and Lovelace's ideas languished in obscurity for a century or more, only to be rediscovered independently by 20th-century computer designers.

Many different ideas and technological advances were necessary steppingstones on the way to the first successful electronic computers. As we have seen, the ability to mass produce standardized precision parts, a hallmark of the Industrial Revolution, was one of these. Meanwhile, the progress of ideas had taken another essential step. In the mid-1800s, George Boole, professor of mathematics at Queen's College in Cork, Ireland, published two works that provided the conceptual foundation for machine logic. In *The Mathematical Analysis of Logic* (1847) and *The Laws of Thought* (1854), Boole explained how fundamental logical processes could be expressed in terms of 1s and 0s by a system now called *Boolean algebra*. (See Sketch 24.) Although Boole reportedly thought his system would never have any practical application, Boolean algebra became the theoretical key to all the "thinking" circuitry of today's computers.

---

[1] The programming language *Ada*, developed in the 1980s, was named in recognition of Lovelace's groundbreaking work.

One more 19th century invention was crucial to the computer revolution. The data from the 1880 U.S. census had taken nearly eight years to process by hand. Herman Hollerith, then a young employee of the Census Bureau, devised a machine that used electricity to automatically sort and tabulate data that had been recorded on punched cards. Hollerith's system reduced processing time for the 1890 census data to only $2\frac{1}{2}$ years. Building on this success, Hollerith founded the Tabulating Machine Company, which eventually became IBM.

The pieces started to come together in 1937, when Claude Shannon, in his Master's thesis at M.I.T., combined Boolean algebra with electrical relays and switching circuits to show how machines can "do" mathematical logic. As World War II spread over the globe, countries on both sides of the battle lines sought military advantage through technological research. Alan Turing, the mathematician who spearheaded the successful British attempt to break the German U-boat command's so-called "Enigma code," designed several electronic machines to help

in the cryptanalysis. Several years earlier, he had conceived the *Turing machine*, a theoretical computer that plays an essential part in determining which kinds of problems are solvable by real computers.

The secrecy surrounding the war effort on both sides, coupled with the urgent need for technological innovations, was the setting for the virtually simultaneous, independent invention of programmable electronic computers by people in several different countries. One of the most interesting instances of this only recently came to light, after nearly half a century of confidentiality. Besides Enigma, the German High Command used a more complex code for many of its transmissions in the later years of the war. Max Newman, a mathematician working at the British code-breaking center at Bletchley Park, devised a way to break that code, but it was far too slow and tedious to be practical. The problem was presented to Tommy Flowers, an electronics engineer working for the British Post Office. In less than a year (from March 1943 to January 1944) Flowers designed and supervised the building of a massive machine that used some 1500 vacuum tubes (like the things in very old radios) to run Newman's decoding process. This machine, dubbed "Colossus," was able to decode German messages in hours, rather than in the weeks or months required to do the same job by hand. Ten of these machines were built and used successfully to decode tens of thousands of German messages, probably shortening the war in Europe by several months and saving thousands of lives. Unfortunately, after the war the British government had all ten machines dismantled and all their technical diagrams burned. The very existence of these machines was kept secret until 1970, and some of their decryption algorithms are still confidential.

Meanwhile in Germany, Konrad Zuse had also built a programmable electronic computer. His work, begun in the late 1930s, resulted in a functional electro-mechanical machine by sometime in the early 1940s, giving him some historical claim to the title of inventor of electronic computers. However, wartime secrecy kept his work hidden, as well. In 1941, Iowa State University professor John Atanasoff and his graduate student, Clifford Berry, built a programmable computer that solved systems of linear equations. The first American general-purpose computer, the *Mark I*, was built by Howard Aiken and a team of IBM engineers at Harvard University in 1944. It used mechanical, electromagnetic relays and got its instructions from a punched paper tape. More than 50 feet long, the Mark I contained about 800,000 parts and more than 500 miles of wire.

February 15, 1946 saw the unveiling of *ENIAC*, (*Electronic Numerical Integrator and Calculator*), built by J. Presper Eckert and John Mauchly of the University of Pennsylvania. ENIAC was intended to help the U.S. war effort by calculating naval artillery firing charts, but the war had already been won without it. It, too, was

ENIAC
(Photo courtesy of IBM Archives)

large: forty-two 9-by-2-by-1-foot panels with more than 18,000 vacuum tubes and 1500 electrical relays, weighing more than 30 tons in all. Its use of vacuum tubes instead of mechanical relays was a major improvement in speed over the Mark I (about 500 times faster) but not in reliability, because vacuum tubes, like light bulbs, burn out with use. Moreover, its programming had to be done manually, by rearranging external wiring and throwing switches, and it had virtually no data storage capacity.

John von Neumann is generally credited with devising a way to store programs inside a computer. His ideas were first implemented in 1949 at Cambridge University on *EDSAC*, (*Electronic Delayed Storage Automatic Computer*). Eckert and Mauchly formed a company that produced and sold the first commercial computer, the *UNIVAC I*

(*UNIVersal Automatic Computer*); it was delivered to the U.S. Census Bureau on March 31, 1951.

The vacuum-tube technology used in these early machines was costly in space, power, and dependability. All that changed when Bell Labs invented the transistor in the early 1950s. Computers using this "second generation" technology became smaller, faster, more powerful, and more economical. The third generation arrived with the introduction of integrated circuitry in the mid-1960s. Personal computers began to become an increasingly affordable reality. As solid-state circuitry was refined and miniaturized, personal computers shrank in size from minis to micros, from desktops to laptops to palmtops. At the same time, computing power, processing speed, and memory capacity increased exponentially.

The first dozen or so years of the 21st century have witnessed a dizzying array of changes in computer technology. Supercomputers can store, analyze, and manipulate massive amounts of data in minutes, providing unprecedented information resources that humankind is just beginning to understand. Miniaturization has led to a symbiosis of cell phones and hand-held computers, promising virtually unlimited communication, knowledge, and entertainment for the price of a good pair of running shoes. The likelihood of additional diverse and exotic innovations to come says that there are more electronic surprises in store for us. A book written in 1990, now ancient by computer-age standards, nevertheless describes well this whirlwind of change:

> There has never been a technology in the history of the world that has progressed as fast as computer technology.... If automotive technology had progressed as fast as computer technology between 1960 and today [1990], the car of today would have an engine less than one tenth of an inch across; the car would get 120,000 miles to a gallon of gas, have a top speed of 240,000 miles per hour, and would cost $4. [2]

For a Closer Look: Much of the information for this sketch is drawn from Ch. 2 of [66]. For the early history of computers, see [9], [68], and [39]. For computing in the 20th century, a good source is [124]. More information about the Colossus may be found at the websites of Pico Technology and the IEEE Computer Society (search for Colossus in each case). See the IBM website for more on the ENIAC and the history of IBM.

---

[2]See [40], p. 17.

The Arithmetic of Reasoning
Boolean Algebra

24

D o computers think? They often appear to. They ask us questions, offer suggestions, correct our grammar, keep track of our finances, and calculate our taxes. Sometimes they seem maddeningly perverse, misunderstanding what we're sure we told them, losing our precious work, or refusing to respond at all to our reasonable requests! However, the more a computer appears to think, the more it is actually a tribute to the thinking ability of humans, who have found ways to express increasingly complex rational activities entirely as strings of 0s and 1s.

Attempts to reduce human reason to mechanical processes date back at least to the logical syllogisms of Aristotle in the 4th century B.C. A more recent instance occurs in the work of the great German mathematician Gottfried Wilhelm Leibniz. One of Leibniz's many achievements was the creation in 1694 of a mechanical calculating device that could add, subtract, multiply, and divide. This machine, called the *Stepped Reckoner*, was an improvement on the first known mechanical adding machine, Blaise Pascal's *Pascaline* of 1642, which could only add and subtract. Unlike the Pascaline, Leibniz's machine used the binary numeration system in its calculation, expressing all numbers as sequences of 1s and 0s.

Leibniz's Stepped Reckoner
(Photo courtesy of IBM Archives)

Leibniz, who also invented calculus independently of Isaac Newton,[1] became intrigued by the vision of a "calculus of logic." He set out to create a universal system of reasoning that would be applicable to all science. He wanted his system to work "mechanically" according

---

[1] See Sketch 30 for more about calculus.

to a simple set of rules for deriving new statements from ones already known, starting with only a few basic logical assumptions. In order to implement this mechanical logic, Leibniz saw that statements would have to be represented symbolically in some way, so he sought to develop a *universal characteristic*, a universal symbolic language of logic. Early plans for this work appeared in his 1666 publication, *De Arte Combinatoria*; much of the rest of his efforts remained unpublished until early in the 20th century. Thus, many of his prophetic insights about abstract relations and an algebra of logic had little continuing influence on the mathematics of the 18th and 19th centuries.

The symbolic treatment of logic as a mathematical system began in earnest in the 19th century with the work of Augustus De Morgan and George Boole, two friends and colleagues who personified success in the face of adversity.

De Morgan was born in Madras, India, blind in one eye. Despite this disability, he graduated with honors from Trinity College in Cambridge and was appointed a professor of mathematics at London University at the age of 22. He was a man of diverse mathematical interests and had a reputation as a brilliant, if somewhat eccentric, teacher. He wrote textbooks and popular articles on logic, algebra, mathematical history, and various other topics and was a co-founder and first president of the London Mathematical Society. De Morgan believed that any separation between mathematics and logic was artificial and harmful, so he worked to put many mathematical concepts on a firmer logical basis and to make logic more mathematical. Perhaps thinking of the handicap of his partial blindness, he summarized his views by saying:

> The two eyes of exact science are mathematics and logic:
> the mathematical sect puts out the logical eye, the logical
> sect puts out the mathematical eye; each believing that it
> can see better with one eye than with two.[2]

The son of an English tradesman, George Boole began life with neither money nor privilege. Nevertheless, he taught himself Greek and Latin and acquired enough of an education to become an elementary-school teacher. Boole was already twenty years old when he began studying mathematics seriously. Just eleven years later he published *The Mathematical Analysis of Logic* (1847), the first of two books that laid the foundation for the numerical and algebraic treatment of logical reasoning. In 1849 he became a professor of mathematics at Queens

---

[2][21], page 331.

College in Dublin. Before his sudden death at the age of 49, Boole had written several other mathematics books that are now regarded as classics. His second, more famous book on logic, *An Investigation of the Laws of Thought* (1854), elaborates and codifies the ideas explored in the 1847 book. This symbolic approach to logic led to the development of *Boolean algebra*, the basis for modern computer logic systems.

The key element of Boole's work was his systematic treatment of statements as objects whose truth values can be combined by logical operations in much the same way as numbers are added or multiplied. For instance, if each of two statements P and Q can be either *true* or *false*, then there are only four possible truth-value cases to consider when they are combined. The fact that the statement "P and Q" is true if and only if both of those statements are true can then be represented by the first table in Display 1. Similarly, the second table in Display 1 captures the fact that "P or Q" is true whenever at least one of the two component statements is true. The final table shows the fact that a statement and its negation always have opposite truth values.

and	T	F
T	T	F
F	F	F

or	T	F
T	T	T
F	T	F

not	
T	F
F	T

Display 1

From there it was an easy step to translate T and F into 1 and 0, and to regard the logical operation tables (shown in Display 2) as the basis for a somewhat unusual but perfectly workable arithmetic system, a system with many of the same algebraic properties as addition, multiplication, and negation of numbers.

and	1	0
1	1	0
0	0	0

or	1	0
1	1	1
0	1	0

not	
1	0
0	1

Display 2

De Morgan, too, was an influential, persuasive proponent of the algebraic treatment of logic. His publications helped to refine, extend, and popularize the system started by Boole. Among De Morgan's many contributions to this field are two laws, now named for him, that capture clearly the symmetric way in which the logical operations *and* and *or* behave with respect to negation:

$$\text{not-}(P \text{ and } Q) \Leftrightarrow (\text{not-}P) \text{ or } (\text{not-}Q)$$
$$\text{not-}(P \text{ or } Q) \Leftrightarrow (\text{not-}P) \text{ and } (\text{not-}Q)$$

Among De Morgan's important contributions to the mathematical theory of logic was his emphasis on relations as objects worthy of detailed study in their own right. Much of that work remained largely unnoticed until it was resurrected and extended by Charles Sanders Peirce in the last quarter of the 19th century. C. S. Peirce, a son of Harvard mathematician and astronomer Benjamin Peirce, made contributions to many areas of mathematics and science. However, as his interests in philosophy and logic became more and more pronounced, his writings migrated in that direction. He distinguished his interest in an algebra of logic from that of his father and other mathematicians by saying that mathematicians want to get to their conclusions as quickly as possible and so are willing to jump over steps when they know where an argument is leading; logicians, on the other hand, want to analyze deductions as carefully as possible, breaking them down into small, simple steps.

This reduction of mathematical reasoning to long strings of tiny, mechanical steps was a critical prerequisite for the "computer age." Twentieth-century advances in the design of electrical devices, along with the translation of 1 and 0 into *on-off* electrical states, led to electronic calculators that more than fulfilled the promise of Leibniz's stepped reckoner. Standard codes for keyboard symbols allowed such machines to read and write words. But it was the work of Boole, De Morgan, C. S. Peirce, and others in transforming reasoning from words to symbols and then to numbers that has led to the modern computer, whose rapid calculation of long strings of 1s and 0s empowers it to "think" its way through the increasingly intricate applications of mathematical logic.

01010100 01101000 01100101   01000101 01101110 01100100

For a Closer Look: A good place to start is pages 1852–1931 of [133]. An accessible book-length treatment is [39]. See also [125] for more on Leibniz, Boole, De Morgan, and C. S. Peirce.

25 Beyond Counting
Infinity and the Theory of Sets

The idea of infinity as an unending process has been a useful mathematical tool for many, many centuries. It is the underlying idea of a limit, the foundational concept for calculus. However, dealing with infinite collections of objects is a relatively new mathematical activity. Only two centuries ago, the great European mathematician Carl Friedrich Gauss said:

> ...I protest above all against the use of an infinite quantity as a *completed* one, which in mathematics is never allowed. The Infinite is only a manner of speaking... [1]

Gauss's comments reflected a common understanding going all the way back to Aristotle. But consider this: We recognize a counting number[2] when we see it, whether it be 5 or 300 or 78,546,291, and we know there is no largest such number because we can always add 1 to any one we have. Now, if we can distinguish the counting numbers from other types of numbers, such as $\frac{1}{3}$ or -17 or $\sqrt{2}$, doesn't it make sense to consider the collection of all counting numbers as a distinct mathematical object? Georg Cantor thought so.

As the American Civil War ended, Georg Cantor was completing his doctorate under the guidance of German mathematician Karl Weierstrass at the University of Berlin. At this time, European mathematicians were in the final stages of tightening up the logical underpinnings of calculus, a process that had been going on for almost 200 years. As they were doing this, they had come to a much deeper understanding of the *real numbers*, the numbers that can be used to label all the points on a coordinate axis. The real numbers can be separated into two distinct types — the *rational* numbers, which can be expressed as ratios of two integers, and the *irrational* numbers, which cannot. Mathematicians knew that each type was "dense" in the other. That is, they knew that

---

[1] Letter to Heinrich Schumacher, July 12, 1831; quoted in [38], p. 120.

[2] Mathematicians usually call these numbers *natural numbers*.

between any two rational points on the number line there are infinitely many irrational points and between any two irrational points there are infinitely many rational points. This had led to a general feeling that the real numbers were evenly divided between rationals and irrationals, more or less.

However, investigations of certain families of functions were beginning to cast doubts on this feeling. Some kinds of functions behaved very differently on these two types of numbers. As Cantor probed these differences, he began to see the importance of considering these various types of numbers as distinct mathematical entities, or "sets." Cantor's concept of *set* was extremely general — and vague:

> By a set we are to understand any collection into a whole of definite and separate objects of our intuition or our thought.[3]

This means that infinite collections of numbers (and other things) can be considered as distinct mathematical objects, to be compared and manipulated just like finite sets. In particular, it makes sense to ask whether or not two infinite sets are "the same size"; that is, whether or not they can be matched in one-to-one correspondence. These elementary ideas rapidly led Cantor to some of the most revolutionary results in the history of mathematical thought. Here are a few of them:

- Not all infinite sets are the same size! (That is, there are infinite sets that *cannot* be put into one-to-one correspondence with each other.)

- The set of irrational numbers is larger than the set of rational numbers.

- The set of counting numbers is the same size as the set of rational numbers.

- The set of all subsets of a set is larger than the set itself.

- The set of points within any interval of the number line, no matter how short, is the same size as the set of all points everywhere on the number line!

- The set of all points in a plane, or in 3-dimensional space, or in n-dimensional space, for any natural number n is the same size as the set of points on a single line!

---

[3]From [24], p. 85, modified slightly to reflect modern terminology.

The abstract simplicity of Cantor's starting point made his set theory applicable throughout mathematics. This made it very difficult to ignore his astounding conclusions, which seemed to defy most mathematicians' common-sense understanding of their subject.

His work was well received in many parts of the mathematical community, but acceptance was by no means universal. Cantor's set-theoretic treatment of infinity generated heated opposition from some of his foremost contemporaries, notably Leopold Kronecker, a prominent professor at the University of Berlin. Kronecker based his approach to mathematics on the premise that a mathematical object does not exist unless it is actually constructible in a finite number of steps. From this point of view, infinite sets do not exist because it is clearly impossible to construct infinitely many elements in a finite number of steps. The natural numbers are "infinite" only in the sense that the finite collection of natural numbers constructed to date may be extended as far as we please; "the set of all natural numbers" is not a legitimate mathematical concept. To Kronecker and those who shared his views, Cantor's work was a dangerous mixture of heresy and alchemy.

To understand what kind of thing Kronecker was worried about, consider the set of all even numbers that can be written as the sum of two odd primes. What numbers are in this set? Well, for any specific even number bigger than four, it's easy (if sometimes tedious) to decide whether or not this number is in the set. For example, 22434 is in the set because $22434 = 12503 + 9931$. Is it true that *every* even number from six on is in the set? We don't know. (That every even number greater than 4 is indeed in this set is a famous conjecture[4] which no one — so far — has been able to prove.) But if we can't say what elements belong to our set, how can we talk of our set as a completed whole? Isn't there a danger that such talk would lead us into contradictions?

Kronecker's fears for the safety of mathematical consistency seemed to be justified by the appearance of several paradoxes in set theory. The most famous of these paradoxes was proposed by Bertrand Russell in 1902. Its set-theoretic formulation need not concern us here;[5] you can get the idea from one of its many popularized versions. Russell himself gave one in 1919:

> A barber in a certain village claims that he shaves all those villagers and only those villagers who do not shave themselves. If his claim is true, does the barber shave himself?

---

[4]It's the Goldbach Conjecture, first proposed by Christian Goldbach in a 1742 letter to Leonhard Euler.

[5]See [43], p. 39, for a more formal, but readable, version of Russell's Paradox.

In slightly more formal terms, is the clean-shaven barber, himself a villager, a member of the set of all villagers who do not shave themselves, or is he not? If he is, then he does not shave himself, but since he shaves all who do not shave themselves, that means he must shave himself, so he isn't in the set, after all. If he isn't in the set, then he does shave himself, but he only shaves those who don't shave themselves, so he must  not shave himself, so he is in the set! There seems to be no way out of this logical loop of contradictions. Dilemmas such as this forced mathematicians of the late 19th and early 20th centuries to undertake a thorough reworking of Cantor's theory of sets in an attempt to free it from the dangers of self-contradiction.

Despite this initial discomfort, Cantor's work has affected mathematics in a decidedly positive way. His basic set theory has provided a simple, unifying approach to many different areas of mathematics, including probability, geometry, and algebra. Moreover, the strange paradoxes encountered in some early extensions of his work encouraged mathematicians to put their logical house in order. Their careful examination of the logical foundations of mathematics has led to many new results and paved the way for even more abstract unifying ideas.

Because much of the opposition to infinite sets was based on philosophical assumptions, Cantor went beyond the usual borders of mathematics to argue for the *philosophical* acceptability of his ideas. He argued that infinite sets were not just interesting mathematical ideas, but that they really did exist. Because of that, his work received attention not just from mathematicians, but also from philosophers and theologians. The time was especially ripe for this, because in the late nineteenth century, just as Cantor's set theory was emerging into the intellectual daylight, an attempt was being made to formulate a philosophy that accommodated both science and religion.

 In 1879, Pope Leo XIII issued the encyclical *Aeterni Patris*, in which he instructed the Catholic Church to revitalize its study of Scholastic philosophy.[6] This type of philosophy is also called Thomism because it is based on the *Summa Theologica* written by Thomas Aquinas in the 13th century. *Aeterni Patris* gave rise to neo-Thomism, a school of philosophical thought that viewed religion and science as compatible. It held that modern science need not lead to atheism

---

[6] An encyclical is a formal letter from the Pope to the bishops which deals with Church doctrine.

and materialism. The neo-Thomists felt that their approach led to an understanding of science that avoided any conflicts with religion (in particular, with Catholicism).

When Cantor's work on the mathematics of infinity became known in the 1880s, it generated considerable interest among neo-Thomist philosophers. Historically, the Catholic Church had held that the claim that infinite things actually exist would lead to pantheism, which was considered a heresy. Cantor, a devout Christian, disagreed. He maintained that his mathematics of infinite sets indeed dealt with reality, but these infinite sets were not to be identified with the infinite God. Cantor distinguished the mathematical aspects of his work from the philosophical ones. In mathematics, he claimed, one is free to consider any concept that is not self-contradictory. Whether there is anything in the real world that corresponds to these concepts is not a mathematical question, but a part of metaphysics.

Metaphysics is the branch of philosophy that studies being and reality. Cantor's metaphysical claim was that actually infinite collections of numbers had a real (though not necessarily material) existence. In patient, persistent correspondence with some of the leading Catholic theologians, he distinguished his position from heretical views and gained semi-official acceptance. Some neo-Thomistic philosophers in Germany even used his theories to assert the existence of actual infinities. For example, they argued that, because the Mind of God is all-knowing, It must know *all* numbers; hence, not only do all the natural numbers actually exist in the mind of God, but so do all rationals, all infinite decimals, and so forth.

The most important effect of set theory in philosophy goes far beyond the arguments of the neo-Thomists, however. The attempts of Cantor and his successors to rid set theory of contradictions and thereby make it metaphysically sound led to deep investigations into the foundations of mathematics. Those investigations early in the 20th century also led to clarifications of logical forms, methods of proof, and errors of syntax, which, in turn, were used to refine the arguments of philosophy. Modern mathematics has provided philosophy with some explicit, formal guidelines for admissible kinds of reasoning and possible logical constructions. Set theory has also provided philosophy with new questions and new ideas about infinity. The boundaries between religion, philosophy, and science have been brought into sharper focus as a result.

One effect of these efforts was that many people came to view mathematics as a subject *removed* from the realm of metaphysics! As people began to work on the foundations of mathematics in earnest, it became clear that the philosophical issues involved were extraordinarily deep. Several schools of thought emerged (see page 55), but none of them found answers that were compelling.

For now, what seems to have won the day is the first part of Cantor's point of view. Mathematics can be done without having to first resolve the philosophical issues. Mathematicians can study infinite sets and, provided they can avoid contradictions, the result will be valid mathematics. Much of this mathematics turns out to be useful in the real world, but the connections are often more subtle and surprising than we might think. Meanwhile, the unresolved philosophical questions (such as the question about *why* mathematics turns out to be applicable) can be left to the philosophers. This might have disappointed Cantor, who cared just as intensely about the metaphysical aspect of his work as he did about the pure mathematics. Most modern mathematicians and philosophers, however, see the recognition of that separation as a giant forward stride in the progress of human thought.

For a Closer Look: Chapter 11 of [48] gives an accessible account of Cantor's work. For a quirky but readable account of ideas of infinity, see [116]. Chapter 2 of [43] is a more general discussion of set theory. Finally, a detailed account of the early history of set theory appears as chapter 5 of [75].

26 Out of the Shadows
The Tangent Function

An ordinary stick was probably the first astronomical instrument. Stuck vertically on a horizontal floor or horizontally into a vertical wall, it will cast a shadow. The length and direction of the shadow reveals the current position of the sun. Observing this variation can tell us what time it is, as in a sundial.

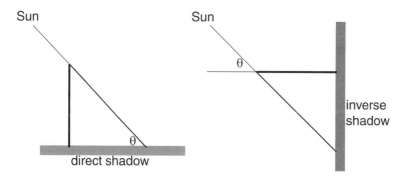

A stick put in either of these positions was called a *gnomon*. With a vertical stick, the shortest shadow will happen at local noon, when the sun is highest, and it was often this shortest shadow that interested early astronomers. The length of the noon shadow depends on the latitude, and it also varies from day to day. It is shortest on the day of the summer solstice, when the sun gets as high up in the sky as it ever does. It is longest on the day of the winter solstice, when the noon sun is lowest. Midway between are the equinoxes, the days when the noon sun is exactly on the celestial equator.

A bit of trigonometry will tell you how the shadow length is related to the sun's angle. Suppose we make the stick one meter long, and we call the angle of the sun θ. Then the shadow of the vertical stick will have length $\cot(\theta)$ meters, and the shadow of the horizontal stick will have length $\tan(\theta)$ meters. In the older texts, the first (our cotangent) was called the "direct shadow" (*umbra recta*), while the second (our tangent) was the "inverse shadow" (*umbra versa*).

At first, trigonometry did not have any such function. Ptolemy's *Almagest*, written around 150 A.D. and the earliest surviving text on trigonometry, used only the chord function, which is closely related to our sine. Nevertheless, Ptolemy explains how to compute the length of the noon shadow at a given latitude at each of the solstices and the equinox. Doing this with the chord table is not hard. No one wants to do these computations over and over, so it's natural to compute them once and put the results in a table.

The mathematicians of medieval India improved on Ptolemy by replacing chords with sines. They, too, computed tables of shadows. The astronomers of medieval Islamic countries who learned trigonometry from India were great makers of tables. A collection of astronomical tables was known as a *zīj*. A *zīj* typically included a table of sines, tables of the position of planets, and lots of other useful data, including tables of shadows.

Surprisingly, the shadow tables were not part of the trigonometric tables, which typically listed only sines. (To find cosines, one would look up the sine of the complementary angle.) In their separate tables, shadows were always given in terms of the position of the sun, rather than just in terms of angles. They were constructed for a gnomon of a certain length, which was usually not the same as the radius used for the sine table.

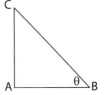

Given a sine table, it was easy to compute the shadow if one knew the sun's angle. In the ancient world, it was always done with ratios:

$$\frac{AC}{AB} = \frac{\sin(\theta)}{\sin(90° - \theta)}$$

What one *cannot* do with such a table, however, is go backwards to find the angle from its cotangent or tangent (as we would call them now). Whenever that was needed, the usual approach was to use the Pythagorean Theorem to find the hypotenuse CB, use that to find the sine of the angle, and then look that up in the sine table. This computation appears over and over in the ancient authors. It's not hard to do, but it does take several extra steps, and all the computations had to be done by hand. A table that allowed them to find the angle corresponding to a given tangent would have saved a lot of work.

What kept Indian and Arabic authors from including the shadows in their trigonometric tables? It seems to have been a kind of conceptual block: Shadows belonged to the theory of sundials and the gnomon, while sine tables were part of trigonometry and astronomy. Shadows were important for determining one's latitude, so they were also part of geography. Each idea stayed very closely attached to the practical context in which it was introduced, and so their relationship was hard to detect.

This began to change in the 11th century, largely because of the combined work of two scholars, Abu Rayhan al-Bīrūnī and Abu-l'Wafa'. Al-Bīrūnī lived most of his life in what is now Uzbekhistan, while Abu-l'Wafa' lived in Baghdad. We know they were in contact with each other because they arranged to observe a lunar eclipse on the same night. By timing the eclipse in each location, they worked out the difference in longitude between their two cities.

The crucial step was taken by Abu-l'Wafa' late in the 10th century, in his book *Almagest*.[1] In it, he introduced all six of the standard functions: sine and cosine, shadow (our cotangent) and inverse shadow (our tangent), hypotenuse of the shadow (our cosecant), and hypotenuse of the inverse shadow (our secant). He figured out how to represent them all on a single diagram, which helped argue for their essential unity.

It is a little less confusing to draw the figure with only three of these functions, as we do here. The point O is the center of the circle whose radius is the one that is to be used for the trigonometric functions. (The current choice, $R = 1$, only became standard much later.) Draw an angle θ, and let C be the corresponding point on the circle, so that the length of OC is the radius R. Drop a vertical line from C to

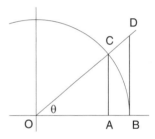

A. Parallel to that, draw the tangent line BD. Then $\sin(\theta) = AC$, $\tan(\theta) = BD$, and $\sec(\theta) = OD$. To add the cosine, cotangent, and cosecant to the picture, all we need to do is draw the equivalent segments for the vertical axis.

Around 1020, al-Bīrūnī wrote an important book called the *Exhaustive Treatise on Shadows*. He included a huge number of topics related to shadows. In particular, he discussed lots of applications of the shadow table and highlighted its usefulness also in the context of trigonometry. Not all mathematicians of the Islamic world adopted the new functions, however. A few of them seemed to prefer to stick to the older system dominated by the sine function, despite the practical advantage of using the shadow function, as well. This version of trigonometry without the full range of trigonometric functions was the one eventually transmitted to Europe. The mathematicians of Europe learned trigonometry from Arabic sources, and at first they used and tabulated only the sine function.

---

[1] We don't know if he chose that name himself; if he did, he must have been supremely confident of its importance!

We can trace how the tangent came out of the shadows in Europe by looking at the work of Regiomontanus (Johannes Müller), who wrote the most important treatise of trigonometry of the 15th century. It was called *De Triangulis Omnimodis* (On All Sorts of Triangles). The idea was to explain how to "solve triangles," that is, how to find the unknown lengths and angles of a triangle from the known ones. Written in the 1450s, *De Triangulis* consolidated most of what was known of trigonometry at its time. But it does not use the tangent function at all.

By 1467, however, when Regiomontanus published his *Tabulae Directionum* (Tables of Directions), he had discovered the usefulness of having a table of tangents, saying that it could "produce great and admirable effects." He called the table *Tabula Fecunda* (Fruitful Table, or Fertile Table), but apparently didn't have a name for the function tabulated there. Some of his followers seem to have called the tangent simply the "fruitful number."

Among the many people influenced by Regiomontanus was Georg Joachim Rheticus, who is best remembered as the person responsible for the publication of Copernicus's *De revolutionibus orbium coelestium* in 1543. While he was working on that book, Rheticus published the trigonometrical sections as a separate book, adding tables that he had computed. Among the innovations found in his tables was the idea of listing the complementary angles on the right side of the page, so that one could use his table of sines to find cosines as well. The resulting little book was published in 1541 and was very popular.

Rheticus spent the rest of his life working, eventually with the help of his student Lucius Valentin Otho, on a more complete book with much more extensive tables. After Rheticus died in 1574, Otho took up the work and completed the book. The *Opus Palatinum de Trianguli* (Imperial Book on Triangles) was finally published in 1596. It is an immense book of 1,400 pages, half of which are dedicated to a huge trigonometric table. It's doubtful that many people read the whole thing, though it is clear that the tables were much used.

In both of his books on trigonometry, Rheticus introduced new ideas and new names. He decided to describe the trigonometric functions in terms of triangles rather than in terms of a circle. Like everyone else of his time, he assumed a fixed radius R for the underlying circle (we would use $R = 1$). The first of his triangles is our standard one: By setting the hypotenuse equal to the radius R, he effectively makes our sine equal to "opposite over hypotenuse" and our cosine equal to "adjacent over hypotenuse." (His sine was R times our sine, of course.) But rather than referring the other functions to the same triangle, he introduced two other triangles in which the radius was in a different position. Here is what it looked like.[2]

If the hypotenuse is the radius of the reference circle, then the base is the cosine and the perpendicular is the sine.

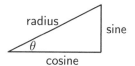

If the base is the radius of the reference circle, then the perpendicular is the tangent and the hypotenuse is the secant.

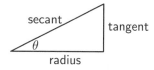

If the perpendicular is the radius of the circle, then the base is the cotangent and the hypotenuse is the cosecant.

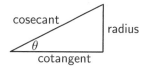

Rheticus's tables reflect this description precisely. He tabulates all six trigonometric functions for all angles from 0° to 45°. He arranges the tables in three columns, each with two subcolumns. The first main column refers to the first kind of triangle, and the subcolumns are headed "perpendiculum" and "basis", i.e., sine and cosine. The second main column is for the second kind of triangle, and the subcolumns are "hypotenusa" and "perpendiculum", secant and tangent. Finally the third column has the third kind of triangle, and the subcolumns are "hypotenusa" and "basis", cosecant and cotangent.

One problem with Rheticus's new names is this name-subname structure. You can't just say "perpendiculum," because that could be either the sine or the tangent. You have to specify the context (the kind of triangle) in each case. On top of that, the choice of which side is the "basis" and which the "perpendiculum" seems arbitrary. It's not surprising that people felt a need for other names. The word "sine" was too established to be changed, but the terms for the other two main functions were still fluid. Was it going to be "umbra," "shadow," or "fruitful number"? As you know, none of those were chosen.

The names we use today were introduced by Danish mathematician Thomas Fincke in a book called *Geometriae Rotundi*. Fincke came up with "rotundi" because he wanted a word that could refer to both circles

---

[2]We follow closely the description in [175].

and spheres; the English translation[3] of the title would be something like "Geometry of Round Things.")

Fincke started from a picture much like the one drawn by Abu-l'Wafa', redrawn here with Fincke's lettering. He saw in the figure the sine (segment UE) and "two other lines associated with it," one tangent to the circle (segment AI) and one secant to it (segment OI). (*Secant* means "cutting"; OI cuts the circle.) So he proposed to call them the *tangent* and the *secant* of θ. "The words, if not the things, are new," he said, "nevertheless they are, we hope, appropriate."

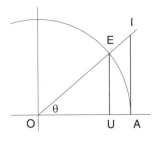

Fincke was somewhat apologetic for introducing new names, but he seems to have felt that Rheticus had opened the door by suggesting his own new names. It seems that he was unaware of "umbra." He dismissed Regiomontanus's suggestion: "For let those people who have called line AI the 'fertile number' look to how they will defend this name; they will not convince me." He pointed out that "perpendiculum" applied to several different lines. And he argued that "Geometry herself has supplied an apt name," since the line in question really was tangent to the circle.

This connection also explains why the secant is the inverse of the *cosine*, rather than the sine. It is the secant that appears naturally in the same picture as the sine and the tangent, with all the "co-functions" appearing from the complementary angle.

The establishment of Fincke's names was not an immediate result of his work. That was the doing of Bartholomew Pitiscus, whose *Trigonometriae* was published in 1595 and quickly became a standard reference. (See p. 153. An enlarged version appeared in 1600 and a translation into English in 1614.) Pitiscus's tables were an improvement on the ones computed by Rheticus, particularly when it came to the values of the tangent and secant of angles close to 90°. And since Pitiscus used the new names introduced by Fincke, those names became accepted as the standard terminology from that time on.

For a Closer Look: The best book on the early history of trigonometry is [175]. It was our main source for this Sketch. There is a short account of the history of trigonometry in [69, ch. 1].

---

[3]Marcos Gouvêa helped us understand Fincke's Latin.

27

Counting Ratios
Logarithms

P ut aside your calculator, take out a large sheet of paper, dip your quill pen into your inkwell, and divide 1867392650183 by 496321087, correct to 10 significant digits. When you're done — perhaps after dinner — you might have some idea of the time-consuming drudgery of arithmetic in the late 16th century. That was a time when European astronomy, impelled by such sharp-eyed observers as Tycho Brahe and Johannes Kepler, required more and more exact calculations to distinguish among competing theories of our planetary system and its relationship to more distant parts of the universe.

Astronomers of that time had extensive tables of sines and cosines, some of which were based on a radius of $10,000,000,000$ to avoid dealing with fractions. As a result, the typical sine was a nine-digit number. (Decimals were not yet in use.) Since many uses of plane and spherical trigonometry require multiplying and dividing sines and cosines, astronomers had to do a lot of tedious arithmetic. Several tricks were used to make the work easier, but they didn't always work well.

Enter the Scottish "laird" John Napier, a landowning gentleman of leisure. In the 1590s, Napier developed an interest in finding ways to simplify computing with large numbers. He was probably inspired by the double sequence that had appeared in Michael Stifel's *Arithmetica Integra* of 1544:

0.	1.	2.	3.	4.	5.	6.	7.	8.
1.	2.	4.	8.	16.	32.	64.	128.	256.

The first of these two sequences is arithmetic — each term differs from its predecessor by the same *amount*. The second sequence is geometric — each term differs from its predecessor by the same *ratio*. Mathematicians of the 16th century did not have exponential notation, but they knew how to calculate with numbers in a geometric sequence by "counting" the number of ratio factors for each one. In Stifel's second sequence, for instance, the ratio is 2, and the number of 2s in each term is counted by the first sequence. To multiply two numbers in that sequence (say 8 and 16), it suffices to add the total number of 2s used for each one ($3+4=7$) and the product will be the number corresponding to that many factors (128). Similarly, division *within the sequence* can be done by subtracting numbers of 2s. Of course, many numbers we might want to multiply or divide are not powers of 2.

Initially, Napier's goal was to extend this computational simplicity to the astronomers' table of sines. For clarity and brevity, we use here some present-day algebraic language and notation to describe his work. Most of that, however, was not part of Napier's toolkit. Working at a time before calculus and coordinate geometry, when even the symbols of algebra were not standardized (see Sketch 8), he described his remarkable insights almost entirely in words.

Napier began with the "total sine," which is $\sin(90°)$, the radius of the implied circle. He visualized it as a segment TS of length $10,000,000$. He envisioned a point g moving from T to S with velocity decreasing so that "in equal times, [it] cutteth off parts continually of the same proportion to the lines [segments] from which they are cut off." For instance, if g moves from T to g_1 and from g_1 to g_2 in equal times, then

$$\frac{Tg_1}{TS} = \frac{g_1 g_2}{g_1 S}$$.

For example, g might move a fifth of the way in the first minute, then a fifth of the remaining segment in the next minute, then a fifth of what was left, and so on. (Napier used $1/1000$ for his computations.) Clearly, g never arrives at S.

Now, here's the brilliant step. Napier envisioned a corresponding point a on an infinite half-line moving with constant velocity in such a way that, as g passed through each g_i point, a passed through the point i. The corresponding positions of a and g at each instant associated each sine with a unique number (generally not an integer) on the half-line. In particular, as a moves through an arithmetic sequence of equally spaced points, the corresponding g-points form a geometric sequence of sines!

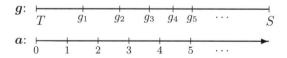

At first Napier called those a-numbers "artificial"; then he decided to dub them "ratio numbers" by putting together the Greek words *logos* (ratio) and *arithmos* (number) to form *logarithm*. Napier defined the logarithm of the segment $g_i T$ to be i multiplied by a large power of 10 (in order to avoid decimals). This makes the logarithm of the total sine 0. Since the point g never arrives at S, Napier said the logarithm of zero was infinite.

In effect, Napier's geometric intuition had extended the idea of exponents from the counting numbers to the entire real line, even though non-integral exponents were not understood at that time. (Napier seems never to have thought about his logarithms as exponents.)

Napier's logarithms had some very nice properties. The key fact was expressed in terms of ratios: if w, x, y, and z are sines such that $\frac{x}{w} = \frac{z}{y}$, then $\log(x) - \log(w) = \log(z) - \log(y)$.[1] This was perfect for applications to trigonometry.

The problem was that this elegant dynamic model did not provide a method for calculating the logarithms of specific sines. Napier tackled this problem with ingenuity and doggedness over a period of many years, and finally published *Mirifici logarithmorum canonis descriptio* ("Description of the marvelous table of logarithms") in 1614. This small book contained 90 pages of logarithmic tables. About 50 more pages described their uses and relations to various geometric theorems, but gave no explanation of how they were constructed. It quickly became a popular tool for astronomers and other scientists. Of course, mathematicians wanted to know how and why it worked. The explanation appeared in *Mirifici logarithmorum canonis constructio*, which only appeared two years after Napier's death in 1617.

In 1615 a copy of the *Descriptio* found its way into the hands of Henry Briggs, professor of geometry at Gresham College in London. Briggs was so impressed that he journeyed to Scotland to visit Napier at Merchiston Tower, his home near Edinburgh. He stayed for a month or so, working with Napier to remedy some difficulties in the system. Two critical changes deserve mention. The first was the choice of "starting point." Both men saw a great advantage in modifying the definition so that $\log(1) = 0$. This, coupled with Napier's result about proportions cited above, leads easily to computing products and quotients via sums and differences of logs. The second, less obvious but equally useful, change was the choice of ratio. Briggs made $\log(10) = 10^{14}$, which allowed his logarithms to work well with our decimal notation for numbers. Once people were more comfortable with decimals, this was simplified to $\log(10) = 1$.

As before, the hard part is computing the actual logarithms, which takes a lot of time. Briggs began by computing (by hand) the square root of 10 to find that $0.5 = \log(3.16227766016837933199\,889354)$. He

---

[1] This generalizes the idea that, if two pairs of terms in a geometric sequence are in proportion, then they must be the same number of steps apart.

computed square roots 53 more times, ending up with a number very close to 1 whose logarithm is

$$2^{-54} = 0.00000\,00000\,00000\,05551\,11512\,31257\,82702\,11815\ldots$$

From that he built back up, using the fact that the logarithm of a product is the sum of the logarithms. In 1624 Briggs published *Arithmetica logarithmica*, a table of the logarithms for the integers 1 to 20,000 and 90,000 to 100,000, accurate to 14 digits. He struggled for the next several years to calculate values for the numbers between 20,000 and 90,000, until a Dutch publisher produced a second edition of Briggs's work (without his knowledge) in which the accuracy was reduced to a mere 10 digits and the missing values were included.

As Napier and Briggs worked in Scotland, the Thirty Years' War was spreading misery on the European continent. One of its side effects was the loss of most copies of a 1620 publication by Joost Bürgi, a Swiss clockmaker. Bürgi had discovered the basic principles of logarithms while assisting the astronomer Johannes Kepler in Prague in 1588, some years before Napier, but his book of tables was not published until six years after Napier's *Descriptio* appeared. When most of the copies disappeared, Bürgi's work faded into obscurity.

While Briggs was calculating, Grégoire de Saint Vincent, a Belgian Jesuit living in Prague, was studying properties of the area under a hyperbola. One of his results, published in 1647, had major implications for logarithms: If $A, B, C, D, \ldots$ is a sequence of points on the x-axis such that the lengths $AB, AC, AD, \ldots$ increase geometrically, then the areas over these segments and under the hyperbola increase arithmetically. As one of Saint Vincent's students observed in 1648, this implies that the areas between the hyperbola and the x-axis segments are logarithms of some sort. Following the terminology of [69], if the hyperbola is the graph of $y = \frac{1}{x}$, we will call the area under it from 1 to some $x \geq 1$ the *hyperbolic logarithm* of x and write it as $\mathrm{Hlog}(x)$.

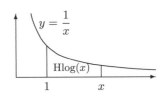

In the 1660s James Gregory, a Scottish mathematician living in Padua in northern Italy, investigated hyperbolic logarithms. Like others of his time, he used large numbers to avoid decimal fractions, which were still relatively unfamiliar despite Napier's extensive use of them. In particular, he used successive approximation with inscribed and circumscribed polygons to calculate $\mathrm{Hlog}(10)$. Adjusted by appropriate powers of 10, his result was

$$\mathrm{Hlog}(10) = 2.302585092994045624017800.$$

This made it clear that hyperbolic logarithms were not the same as the ones developed by Napier and Briggs. It also showed that a logarithmic relationship may appear "naturally" when computing areas.

As Gregory was working on Hlog in Italy, Isaac Newton in England was investigating area under the hyperbola $y = \frac{1}{1+x}$. Between 1664 and 1665, Newton determined that the area under this hyperbola from 0 to some $0 < x < 1$ can be calculated by the infinite series

$$x - \frac{x^2}{2} + \frac{x^3}{3} - \frac{x^4}{4} + \frac{x^5}{5} - \frac{x^6}{6} + \cdots$$

Since this hyperbola is just a translation of $y = \frac{1}{x}$ one unit to the left, this series represents $\text{Hlog}(1 + x)$ — in our notation, but not in his.

About ten years later, Newton provided one more piece in the Hlog puzzle, albeit indirectly. By that time, integral exponents had been in general use for awhile, as a shorthand for repeated multiplication. Fractional powers, on the other hand, were not commonly used, and this kept Napier and others from thinking of their logarithms in terms of exponents. Around 1675, Newton sent a letter to the Royal Society of London explaining some of his work on infinite series, including his binomial theorem, which gave a series for powers of $(1 + x)$. Because this worked for fractional powers too, Newton described what they meant. This unlocked the door to a more general understanding of the exponentiation process, a door opened wide in the following century.

A systematic description of logarithms as exponents was published in 1742 by William Gardiner, citing William Jones as a source of his work. But what really established this point of view was the authority of the eminent Swiss mathematician Leonhard Euler. His two-volume *Introductio in Analysin Infinitorum*, the most influential mathematics text of its time, was published in 1748. In it he describes exponentiation as a function $y = a^z$ for any real number z, presuming without explicit definition that the idea of fractional exponents can be extended to all real numbers. He then says:

> "[C]onversely, given any positive value of y, there is a convenient value of z that makes $a^z = y$; this value of z ... is usually called LOGARITHM of y."

But which logarithm? In Euler's approach there is a log function for each positive base $a \neq 1$. Euler showed that all logarithms of a number y are multiples of each other. Specifically (in modern terms), for any two bases a and b, there exists a number K (depending on a and b, of course) such that

$$\log_b(y) = K \log_a(y).$$

So which log function is "best"? Is it the base-10 logarithm of Briggs, or Hlog, or something else? To a certain extent, the answer depends on what you're doing. Certainly, Briggsian logarithms are very convenient for making tables and for calculating with our decimal numeration system. But is that the most convenient base for the mathematical theory? Euler resolved that question using infinite series. The main idea works like this.[2]

For any logarithm, the base is the number whose log is 1. Now consider $y = a^z$ for some $a > 1$. Euler used Newton's binomial theorem to expand a^z in an infinite series involving a constant k whose value depended on the value of a:

$$a^z = 1 + \frac{kz}{1} + \frac{k^2 z^2}{1 \cdot 2} + \frac{k^3 z^3}{1 \cdot 2 \cdot 3} + \frac{k^4 z^4}{1 \cdot 2 \cdot 3 \cdot 4} + \cdots$$

If we choose a "nice" value for a, say $a = 2$, the constant k turns out to be a very ugly number. Euler decided that it would be best to choose the constant a so that the annoying k would turn out to be equal to 1, making the formula very simple. Euler dubbed this base e (the first letter of "exponential"); so we get

$$e^z = 1 + \frac{z}{1} + \frac{z^2}{1 \cdot 2} + \frac{z^3}{1 \cdot 2 \cdot 3} + \frac{z^4}{1 \cdot 2 \cdot 3 \cdot 4} + \cdots$$

When Euler computed the logarithm function for his carefully chosen base e, he got

$$l(1 + x) = x - \frac{x^2}{2} + \frac{x^3}{3} - \frac{x^4}{4} + \cdots,$$

which is the series that Newton had found for Hlog! Thus, the simplest member of the family of log functions is actually the hyperbolic log, known now by Euler's name for it, *natural logarithm*, and sometimes written "$\ln(x)$".

Finally, using his series expansion, Euler calculated the derivative of $f(x) = \ln(x)$, and saw that $f'(x) = \frac{1}{x}$. Of all the log functions, this one has the simplest derivative, making it truly deserving of the name "natural."

For a Closer Look: The information for this Sketch was drawn primarily from Chapter 2 of [69] (pp. 68–147). A shorter and very clear account of Napier's work can be found in section 13.2 of [99]. Many sourcebooks provide extracts from the relevant texts.

---

[2]See pp. 139–141 of [69] for a fuller explanation.

28

Any Way You Slice It
Conic Sections

This is a tale of how pursuit of a problem posed by a Greek oracle
unlocked a secret of the solar system and helped to start the
Scientific Revolution. It is a case study in how a mathematical
tool, once discovered, can prove useful in unexpected ways.

A legend from ancient Greece tells how the people of the island of
Delos, afflicted by a plague, appealed to the oracle of Apollo. The oracle
told them to double the size of Apollo's cubical altar. They built a new
altar — some say by putting two cubes together, others say by doubling
the length of each side. The plague continued, and the Deliansrealized
that the oracle wanted a cubical altar of exactly twice the *volume* of
the original. When they turned to Plato for advice, he told them that
"by this oracle [Apollo] enjoined all the Greeks to leave off war and
contention, and apply themselves to study, and... to live peaceably with
one another, and profit the community."[1] That is how the problem of
"duplicating the cube" — finding a geometric construction for what we
would call $\sqrt[3]{2}$ — became known as the Delian Problem.

Greek mathematicians attacked this problem in various ways. Before Plato's time, Hippocrates of Chios had shown how the Delian Problem reduces to finding line segments that form two mean proportionals.
Specifically, if a is the side length of the original altar and if there are
segments of lengths x and y such that the three ratios $a : x$, $x : y$, and
$y : 2a$ are equal, then x will be the side length of the doubled altar.

The next step came half a century later. In the last half of the
4th century B.C., the Macedonians under Alexander conquered all the
eastern Mediterranean from Greece to Egypt and as far east as central
India. Many of those regions came to share in the unifying influence of
a common language, Greek. Many ancient mathematicians we now call
Greek came from various parts of that sprawling empire. Menaechmus,
from an area that is now part of Turkey, was one of several mathematicians to solve the problem of finding two mean proportionals. He did
it using curves obtained from three-dimensional geometry, by slicing a
cone. Each of the curves represented the equality of two of the ratios,
so intersecting two of them yielded the desired line segment.

---

[1]This version of the story is told by Plutarch in *De Genio Socrates* 579. See
[112, p. 399]

The person most closely tied to curves formed by slicing cones is Apollonius of Perga (in Turkey), the "Great Geometer" of the third century B.C., who built on the ideas of Menaechmus, Aristaeus, and Euclid to unify these curves in a single, elegant theory. Apollonius began with a double cone formed by rotating two intersecting lines around an axis that bisects their angle of intersection. When the cone is cut by a plane that does not pass through the tip of the cones, the points of intersection on the plane form a curve called a *conic section*, or simply a *conic*.

By rotating the plane so that its angle with the axis changed from perpendicular to parallel, Apollonius produced all four kinds of conic sections — circle, ellipse, parabola, and hyperbola. The last three of these names were derived from earlier work on properties of areas. Loosely translated, they imply "too little," "just right," and "too much."[2] (Shades of Goldilocks!)

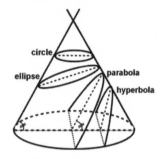

Here is why Apollonius chose those names. For each type of curve, he looked at the line segment formed by arbitarily choosing a point P on it and dropping a perpendicular to the major axis. He compared the area of a square built on that segment with a rectangle formed by

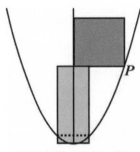

Parabola: areas equal

a segment equal in length to what we now call the *latus rectum* and the segment of the major axis determined by the end of the perpendicular from P and the nearest vertex of the curve. If the areas are exactly equal, the curve is a parabola; if the square area is too small or too large, the curve is an ellipse or a hyperbola, respectively, and the exact shape could be described in terms of how much smaller or larger the area was.

Apollonius's *Conics* was a long, detailed account of conic sections and their uses. It encompassed earlier work, but it went far beyond what others had done. During the following several centuries of Greek scholarship, others expanded the theory of conics and applied it to various problems. Notable among them was Pappus of Alexandria, a mathematician of the early 4th century A.D. who used properties of the

---

[2]For more on the etymology of these names, see [10, Capsule 61].

focus and directrix of a hyperbola to find a method for trisecting an angle.[3] But Apollonius's *Conics* remained the definitive work on conic sections for a very long time. Four of its eight books have survived directly from the Greek versions, and three others have come down to us by way of translations from Arabic; one book has been lost.

It is important to keep in mind that the Greeks' treatment of conics, and of geometry in general, was entirely "synthetic." That is, it dealt with lines, squares, and various other shapes, not with numbers or coordinates. The algebra and coordinate geometry that we take for granted were nearly 2000 years in the future. Comparisons of lengths, areas, and the like involved proportionality, not explicit measurement. In hindsight, one can see in Apollonius's work some powerful antecedents of a coordinate treatment of conics, but the machinery of analytic geometry simply didn't exist then.

In the several centuries following the collapse of the Graeco-Roman Empire and the spread of Islam around the Mediterranean Sea and eastward, many Greek mathematical writings were translated into Arabic. In the 9th and 10th centuries, Arab[4] and Persian mathematicians began to develop the algebra of equations. This early algebra, done entirely in words, began by listing and solving quadratic equations. The next step was the search for ways to solve cubic equations (which are, in fact, extensions of the Delian Problem). A prominent player in this game was 'Umar al-Khāyammī,[5] a 12th-century Persian scholar whose writings spanned poetry, mathematics, science, and philosophy. Building on the work of several Arab mathematicians of the previous century, al-Khāyammī solved several different forms of cubic equations using the method of Menaechmus, i.e., by intersecting two conic sections.

There is some evidence that Apollonius's *Conics* was known in Europe as far back as the 13th century, when Erazmus Witelo used conics in his book on optics and perspective. Nevertheless, there was little new work on the subject for a long time. Perhaps it was simply a matter of Apollonius being hard to read and giving the impression that he had basically done it all already.

The Renaissance of European art and learning began in earnest during the 15th century. When Constantinople fell in 1453, some Greek scholars brought their manuscripts and knowledge to the cities and new

---

[3]This was another of the three "Problems of Antiquity," closely related to the Delian Problem. See page 21.

[4]As it was with "Greek", the term "Arab" refers to someone who lived somewhere in the Islamic Empire and wrote in Arabic.

[5]Perhaps better known these days as Omar Khayyám, author of the *Rubáiyát*.

universities of the West. Movable-type printing was invented, making
it much easier to disseminate new ideas. And there were lots of new
ideas, including Nicolaus Copernicus's new theory of the solar system,
published early in the 16th century. This was a watershed period for
early European science and mathematics. Algebra was becoming systematically symbolic (though not yet standardized) and decimal representation of fractions was spreading rapidly in the scientific community.
Napier was developing logarithms (see Sketch 27) and Galileo was using
the parabola to describe projectile motion.

German astronomer Johannes Kepler was living in Graz in the summer of 1600, when there was a solar eclipse. He assembled in the city
square a large wooden instrument that he used to observe the eclipse.
It was a kind of large-size pinhole camera, designed to allow him to
see and measure what was happening without hurting his eyes. Now,
light can do funny things when it goes through a pinhole, and Kepler's
observation led him to wonder about some questions of optics. After
all, astronomical observations were all done by looking at the sky, and
weird optical effects might affect the precision of those observations.

Soon after that, Kepler had to leave Graz. He ended up in Prague,
where he worked at the Imperial Observatory. At first he was Tycho
Brahe's assistant, but Brahe died in 1601 and Kepler inherited both
his job and an immense amount of observational data that Brahe had
collected over many decades. Using this scientific treasure-trove, he set
out to try to work out the orbits of the planets. But that work went
slowly, and he decided to write out his work on optics first.

He started by reading a book that contained two famous works on
optics, one by Ibn al-Haytham (also known as Alhazen) and the other
by Witelo. Perhaps it was because they both quoted Apollonius that
Kepler decided he needed to read that, as well. Luckily, a good Latin
translation by Commandino had been published in 1566. It was hard
work. In 1603, he wrote to a friend that "All the *Conics* of Apollonius
had to be devoured first, a job which I have now nearly finished."[6]
Kepler's *Optics* was finally published in 1604.

They say that chance favors the prepared mind. As Kepler was
doing all this, he was also studying the planet Mars. A convinced
Copernican, Kepler assumed at first, like everyone else, that the orbit
had to be a circle with the Sun somewhere near the center. But after
examining Tycho Brahe's observations, Kepler realized that the orbit
of Mars did not fit a circular pattern: it was flatter than a circle, some
kind of oval. Of course, he thought of the conic sections! By further

---

[6]Cited in [103, p. xii].

careful observation and measurement, he concluded that the orbit must be an ellipse with the sun at one focus. He asserted that all planets have such orbits and set out three "laws" that described the orbits in detail. He also argued that there must be some sort of force exerted by the Sun on the planets that caused them to move along such orbits.

The 17th century was a time of great progress in understanding motion. As a result, it became possible to work out whether a central force would produce orbits that followed Kepler's laws. Kepler had suggested that the force would get smaller with distance. By the 1680s, the more common guess was that the force should get smaller with the *square* of the distance, but no one knew how to relate the nature of the force to the shape of the orbit.

In 1684, British astronomer Edmund Halley visited Isaac Newton to discuss this problem. Newton told him that he knew that an inverse-square central force would lead to Kepler's elliptical orbits. Halley was surprised and asked him how he new. Newton's answer was simply "I have calculated it." It was Halley's request that Newton write up this calculation that led, after a few years, to the publication of Newton's masterpiece, *Philosophiae Naturalis Principia Mathematica*[7] Kepler's laws were thereby confirmed, and thus the conic sections were inscribed in the heavens!

Of course, there had been progress between 1600 and 1680. Mostly, it had to do with two radically new methods. The first was the coordinate geometry invented by Descartes and Fermat. (See Sketch 16.) As Fermat showed, every equation of degree 2 in two variables describes a conic section. His proof boils down to showing that the curve defined by such an equation always had one of the properties that Apollonius called the "symptoms" of the conics.

The other new method was projective geometry. Girard Desargues and Blaise Pascal extended the principles of perspective drawing to create a new geometry. In Euclidean geometry, two figures are congruent if one is the rigid-motion image 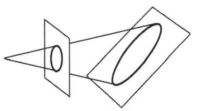 of the other. In projective geometry, one figure is allowed to be transformed into another by a "projection." Think of images projected onto a screen from a light source passing through a film. If a circle is on the film, the rays passing through the circle from the point of light form

---

[7]Literally, "mathematical principles of natural philosophy." What Newton's title is saying is that he has developed a *mathematical* theory of the physics of motion.

a cone. If the screen is parallel to the film, the image is a (bigger) circle. As you tip the screen, the image changes into an ellipse, then a parabola, then a hyperbola. Thus, in projective geometry, all four kinds of conic sections are essentially "the same" because you can get from one to another by a projective transformation. (See Sketch 20.)

Coordinate geometry and projective geometry come together in what became known as "algebraic geometry," the geometry of figures defined by polynomial equations. The conic sections — the family of all curves described by two-variable polynomials of degree 2 — are the simplest algebraic curves. They serve as model, motivation, and source of questions for the study of other algebraic curves and surfaces.

The modern world has seen increasingly sophisticated applications of conic sections — parabolic reflectors and optical lenses, elliptical satellite orbits, hyperbolic radio-wave navigation — undoubtedly with more to come. Amid that impressive diversity, there is a striking unity to this powerful quartet of curve families. Emerging from a single problem, they are unified by the geometry of their constructions and the algebra of their descriptions. Moreover, their utility is a testament to the value of pure curiosity. In the words of the 19th-century British mathematician J. J. Sylvester,

> "But for this discovery [of conic sections], which was probably regarded in Plato's time and long after him, as the unprofitable amusement of the speculative brain, the whole course of practical philosophy of the present day... might have run in a different channel; and the greatest discovery that has ever been made in the history of the world, the law of universal gravitation, ... might never to this hour have been elicited."[8]

For a Closer Look: Apollonius's *Conics* and Kepler's *Optics* are both available in English translations. Coolidge's [31] is showing its age, but remains a good source of references to original work. All the textbooks include discussions of the ancient work, but tend to get thinner on the later material.

---

[8]from "A Probationary Lecture on Geometry," [173, vol. 2, p. 7]

29 Beyond the Pale
Irrational Numbers

L egend has it that Pythagoras became fascinated with numbers by listening to music. If you pluck a taut string on a lyre or a guitar, you hear a musical note. If you pluck two strings at once, the notes may or may not sound good together. It is said that Pythagoras noticed how the harmonic quality of the notes depends on the ratio of the lengths of the strings. When it is the ratio of two small whole numbers, the strings sound good together. For example, if one string is twice the length of the other, a 2 : 1 ratio, the notes are an octave apart; if the ratio is 3 : 2, the notes form a "perfect fifth." But if the ratio is something like 11 : 8, the notes are dissonant. Having realized this, the Pythagoreans[1] put a lot of effort into understanding ratios. And they made an amazing discovery: there are lengths whose ratios cannot be expressed by numbers at all!

One such ratio relates the diagonal of a square to its side. No one really knows how the Pythagoreans' proof went, but Aristotle says that it involved a contradiction based on even and odd numbers. Here, in modern notation, is an ancient proof like that:

Start with a 1-by-1 square, and suppose that the ratio of its diagonal d to its side is $a : b$, where a and b are whole numbers. Then d can be written as a fraction $\frac{a}{b}$ in lowest terms. Since a and b have no common factors, at least one of them must be odd. Now, by the Pythagorean Theorem,

$$\left(\frac{a}{b}\right)^2 = 1^2 + 1^2, \quad \text{so} \quad \frac{a^2}{b^2} = 2, \quad \text{implying} \quad a^2 = 2b^2.$$

This means that a^2 is even, so a must be even. (If a were odd, a^2 would be odd as well.) Since a is even, we can write it as $2s$, where s is some other whole number. Then, substituting $2s$ for a, we have

$$(2s)^2 = 2b^2, \quad \text{or} \quad 4s^2 = 2b^2, \quad \text{so} \quad 2s^2 = b^2,$$

implying that b^2 is even, so b must be even. So a and b are both even! That can't be, as we saw, so the only logical escape is to recognize the original hypothesis as false. That is, the ratio $d : 1$ *cannot* be expressed as the ratio of whole numbers.

---

[1]For more about Pythagoras and the Pythagoreans, see page 17.

Today we would say that $d = \sqrt{2}$ and call it an *irrational num-ber*, but the Pythagoreans called the two segments "incommensurable" (cannot be measured together) because there is no common unit of measurement for which d is a units and 1 is b units. The ratio itself they called "irrational" or "inexpressible."

Similar arguments easily produce other pairs of incommensurable segments. Greek mathematicians absorbed the discovery well. They concluded that numbers and segments were two dramatically different kinds of quantities. Numbers belonged in arithmetic, segments in ge-ometry. The link between them was the notion of ratio, but ratios of segments were far more complicated than ratios of numbers. A theory of ratios of general quantities was eventually created, probably by Eu-doxus, a physician and legislator who was a pupil of Plato, in the 4th century B.C. It is the content of Book V of Euclid's *Elements*.

Incommensurability continued to be a productive irritant in Greek and medieval European thought for quite a long time. Indian and Arab mathematicians, on the other hand, had no qualms about dealing with "surds" (as roots of rationals were often called). They simply accepted the notion that there must be numbers whose squares were 2 or 3 or 5, etc., and set about finding ways to work with them. For instance, in 9th-century Egypt, Abu Kamil worked with surds of all sorts, both as coefficients and as solutions to algebraic problems. In Baghdad around 1000 A.D., Abū Bakr al-Karajī wrote a book on arithmetic and algebra in which he described how the Greeks' geometric irrationals could be treated as numbers. In 11th-century Persia (now Iran) the poet/astronomer Omar Khayyám examined Eudoxus' theory of pro-portions from a numerical viewpoint that was well ahead of its time. And in 12th-century India, Bhāskara II[2] described rules for calculating with square roots of non-square integers.

All of this knowledge was transmitted to Europe, initially through contact with Arab culture and then by examining Greek manuscripts preserved in Byzantium. Europeans thereby inherited two different points of view. One, from the Greeks, treated numbers and magnitudes as very different and connected only by the notion of ratio. The other, from the Arabs, used a broader concept of number that included whole numbers, fractions, and various kinds of roots.

Unification of these two viewpoints came by way of decimal frac-tions, as in Simon Stevin's book, *The Tenth*.[3] In another book, Stevin said explicitly that rationals and irrationals alike are equally entitled to

---

[2]See page 26 for the distinction between Bhāskara I and Bhāskara II.

[3]See Sketch 4 for more about this.

be called numbers, and his approach showed that they could all be ordered in a way that we now call the "number line." Decimal fractions greatly simplified calculation with fractions and roots, so they were quickly accepted by scientists and engineers throughout Europe. Thus, when Descartes coordinatized the plane in *La Géométrie* of 1637, his assertion that all the points on his axis corresponded to *real numbers* (as he named them) was not controversial. The fact that some were rational and some were not was irrelevant to his main ideas and to the subsequent mathematical advances of the late 17th and early 18th centuries, most notably the methods of calculus. The question of rationality versus irrationality retired to the rooms reserved for abstract mathematics, often ignored but never forgotten.

For instance, at that time it wasn't clear whether or not π is rational. It surely isn't exactly $\frac{22}{7}$ or even $\frac{355}{133}$, but might it not be some exotic quotient of gigantic integers? The question remained open until 1761, when Swiss mathematician/physicist Johann Lambert proved that π is irrational. Almost two decades later his famous compatriot, Leonhard Euler, proved the irrationality of e, the natural base of logarithms. Investigations of these and other specific irrationals appeared in European mathematical literature during the 18th and 19th centuries.

How many of these irrational numbers existed was also not clear. The first irrationals discovered were n-th roots of rational numbers. All such numbers are solutions to polynomial equations with integer coefficients; they are called *algebraic* numbers. A simple example is $\sqrt{2}$, which is a solution to $x^2 - 2 = 0$. So are the roots of $2x^5 - 4x + 1 = 0$, which are far more complicated. Are *all* irrational numbers algebraic? The existence of some that are not was suspected as early as late 17th century, when John Wallis conjectured that π was such a number. Such numbers had a collective name — *transcendental* — but their existence was unproven until the 19th century.

Verifying that a number is transcendental means proving that there is *no* polynomial equation with integer coefficients for which that number is a solution. That's hard. The first provably transcendental number, constructed in 1851 by Joseph Liouville of France, was

$$\sum_{n=1}^{\infty} \frac{1}{10^{n!}} = 0.11000100000000000000000010\ldots.$$

(the 1s appear at positions 1!, 2!, 3!, etc.) The "obvious" candidates were stubbornly elusive. Finally, in 1873, the French mathematician Charles Hermite succeeded in proving that e^r is transcendental for all rational numbers r, but predicted that π would be a much more difficult challenge. That challenge was met nine years later by Ferdinand Lindemann of Germany.

Meanwhile, more and more students in engineering and science were learning the powerful methods of calculus that had been developed during the 18th century. Some of their teachers — notably Augustin Louis Cauchy in France and Richard Dedekind in Germany — became increasingly uncomfortable as they tried to explain the slippery concept of the real-number continuum to those students. They began to search in earnest for a more precise description of the real numbers.

In the last half of the 19th century, two very different, equally elegant models of the continuum appeared in Germany almost simultaneously. Building on Cauchy's work with sequences, Georg Cantor approached the question arithmetically, starting with decimal expressions of rationals. Richard Dedekind, on the other hand, took a fundamentally geometric approach. The next few paragraphs present a broad-brush outline of the main ideas of both theories. To do that, we authors must engage in an almost embarrassing amount of printed hand-waving, ignoring or finessing many significant logical issues and hoping that thereby the big pictures will emerge more clearly.

Cantor, building on the work of Karl Weierstrass, began by observing that any point on the line can be approximated from below to within any desired degree of accuracy by an increasing sequence of decimal fractions. For instance, the sequence $1, 1.3, 1.33, 1.333, \ldots$ approaches the point labeled $\frac{4}{3}$, which is called its *limit*. The shorthand for this sequence is the infinite decimal $1.333\ldots$. A similar sequence is determined by the point $\sqrt{2}$, which is between 1 and 2. Squaring shows that it is between 1.4 and 1.5, between 1.41 and 1.42, and so on. This yields the sequence $1, 1.4, 1.41, 1.414, 1.4142, \ldots$; it can approach $\sqrt{2}$ as closely as your computational perseverance permits. The visual intuition that any point can be trapped uniquely within successive powers-of-tenths intervals survives a more formal logical treatment. That is, choosing points for 0 and 1 determines a one-to-one correspondence between *all* the points on the line and all infinite decimals — well, almost. Some fussy special cases need attention, but they can be handled. With a bit of permissible logical tinkering, this approach leads to a representation of the real numbers as the set of infinite decimals.

This is good news and bad news. The good news: We now have a unified representation of all real numbers, rational and irrational alike, and we can tell which is which from their decimal expressions. It is not hard to show that the decimal for any rational number must eventually have a finite repeating pattern, so all the other infinite decimals represent irrationals. The bad news: Doing arithmetic with infinite decimals is pretty much impossible. Worse, treating an infinite sequence as an actually infinite "thing," not merely a potentially unending process, was unacceptable to many mathematicians of that day. The time of Cantor's set theory had not yet come (but would soon), and the idea of a number system as a set of objects with well-behaved, but abstract, arithmetic operations was just beginning to take shape.

In an 1872 paper, Cantor generalized this model to encompass limits of all rational-number sequences that "ought to" have them. Almost half a century earlier, Cauchy had described such sequences (now commonly called *Cauchy sequences*) by means of a convergence test that did not presuppose the existence of a limit.[4] Defining two such sequences *equivalent* if their difference sequence converges to 0 creates equivalence classes in which all the sequences in a particular class "ought to" have the same limit. By describing ways to add and multiply them, Cantor turned that set of equivalence classes into a model of the real number system. Moreover, since each class contains an infinite-decimal sequence, the infinite decimals can still be used to represent them.

Dedekind's geometric approach focused on the points of the number line that don't have rational labels. He observed that each point separates the rationals into two disjoint sets — all rationals less than the point, and all the rest. Since there is a rational number in any interval of the number line, no matter how small, different separating points must determine different pairs of sets. If the chosen point is rational, then that number is the least element of the "upper" set. If there is no least rational in the upper set, that pair must correspond to an irrational point. Dedekind defined ways to add and multiply these pairs so that they behaved like numbers. This made the pairs of sets, now known as *Dedekind cuts*, a model of the real numbers.

These formal representations of the continuum did not sit well with some contemporary mathematicians. One of the most outspoken critics was Leopold Kronecker, a prominent professor at the University of Berlin, who declared that "irrational numbers do not exist."[5] For measurement, he had a point. The difference between any irrational and a nearby rational number can be made as small as we please, so rational approximations are always good enough for such purposes. (Using 3.14159265 instead of π in calculating the circumference of the Earth makes a difference of about 2 inches.) But while the rationals may be sufficient for calculation, it is the irrationals that provide the conceptual richness of continuity that underlies calculus and analysis.

---

[4]A sequence $\{a_n\}$ is *Cauchy* if, for any $\varepsilon > 0$, there is some natural number N such that, for all $m, n > N$, $|a_m - a_n| < \varepsilon$.

[5]Stated in a letter to Ferdinand Lindemann; see p. 204 of [84].

At the root of Kronecker's displeasure was the fact that both constructions regarded infinite sets as complete things, rather than simply
as processes that could be continued at will. Dedekind's cuts were actual sets. Cantor's equivalence classes of sequences were infinite sets
as well. Moreover, both treated the set of all such objects — the real
numbers — as an object itself. Kronecker, who did not like any sort of
actual infinity, regarded talk of an infinite collection of infinite collections as sheer lunacy.

Once Cantor had crossed that barrier and started to think about
the real numbers as a whole, he found some astonishing things. Using
one-to-one correspondence to define "being of the same size," he showed
that there are different sizes of infinite sets. He showed that the set
of all real numbers is larger than the set of rationals. In other words,
there are far more irrational numbers than rational ones. This may not
seem so surprising to you. After all (you might say to yourself), most
roots of rationals are irrational, and there are lots of those. But here's
the next surprise: Dedekind pointed out to Cantor that the set of all
algebraic numbers is the same size as the set of rationals. This means
that *most* irrational numbers are transcendental!

Kronecker's skepticism notwithstanding, the real numbers are so
useful that today everyone accepts them. Many mathematicians have
continued to investigate transcendental numbers. Their quest was encouraged by David Hilbert's famous 23-problem challenge for the 20th
century, delivered in his address to the Second International Congress
of Mathematicians in 1900. Problem 7 asked what could be said about
numbers of the form α^β, where α is algebraic (and not 0 or 1) and β
is an algebraic irrational. An example is $2^{\sqrt{3}}$. In 1934, both Aleksandr
Gelfond of Russia and Theodor Schneider of Germany proved that *all*
such numbers are transcendental.

Despite these and other impressive results, membership in the pro-
fusion of transcendentals remains largely mysterious. Even such simple,
inviting combinations as $\pi + e$, πe, and π^e are still among the "unde-
cided." We know that most real numbers are irrational, and most
irrational numbers are transcendental. But we don't know what most
transcendentals look like. There's a lot more to be done.

For a Closer Look: Julian Havil's [84] gives an accessible account
of irrational numbers and their history. The best description of Dedekind's "cuts" is the original one, which makes up the first half of [41].
You can read about Hilbert's problems and their solvers in [178]

30 Barely Touching
From Tangents to Derivatives

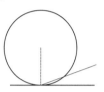

It's one of the standard problems in any calculus text: find the tangent line to the curve $y = f(x)$ at $x = a$. One might think that derivatives were invented precisely to solve this problem. But finding tangent lines is something that goes back long before calculus.

A theorem in Euclid's *Elements* says that if we take a point on a circle and draw a line through that point perpendicular to the radius, the line will "touch the circle." In Latin, "to touch" is *tangere*, so the touching line was "tangente," whence *tangent*. Euclid went on to say that, if you try to fit another line in between the tangent line and the circle, you will fail, since that other line would actually cut the circle at a second point.

The Greeks studied many other curves besides the circle. In particular, they spent a lot of time studying the conic sections. (See Sketch 28.) In Apollonius's *Conics*, there are explanations of how to find tangent lines to hyperbolas, ellipses, and parabolas.

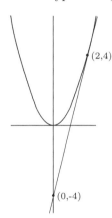

Take the case of the parabola. It is no longer right to say that a line that intersects the curve only once is tangent to it. Any vertical line intersects the parabola $y = x^2$ only once, but it's certainly not a tangent. So the first thing we need is to explain what a tangent line actually is. Apollonius defined it, following Euclid: The line is tangent if it is impossible to fit another line into the angle it makes with the parabola. Of course, Apollonius did not use an equation to describe the parabola, but we can translate his theorem for our standard parabola $y = x^2$. Here's what it says. Take $x = a$, so that $y = a^2$. Find the point on the y-axis that is on the other side of the vertex and at the same distance, i.e., the point $(0, -a^2)$. Draw the line from (a, a^2) to $(0, -a^2)$; it will be tangent to the parabola. A quick calculation using derivatives checks that this is indeed correct.

The conic sections being done, tangent-finding seemed to become less interesting. To have new problems, one needs new curves. A few were known and studied, but it seems that the problem of their tangents was either not considered or turned out to be too hard. In any case, the question had to wait until new curves became abundant, which happened in the 17th century.

Given their interest in all things Greek, mathematicians of early modern times knew about the conic sections and their tangents. They were also interested in how things move. Thinking about motion led to a new and interesting curve. Suppose a circle is rolling along a straight line. Its center just moves forward, but what curve does a point on the circumference trace out? The question seems to have been asked first by Nicholas of Cusa. Marin Mersenne made a more precise description of the curve, known as the *cycloid*, and popularized the obvious questions: What is the area under it? What is its length? Can we find tangent lines?

The cycloid apparently fired the imaginations of geometers everywhere, because everyone worked on it. Galileo, Mersenne, Roberval, Wren, Fermat, Pascal, and Wallis all made discoveries and then fought about who had done it first. The real action, however, started after Fermat and Descartes invented coordinate geometry. Their work vastly expanded the range of available curves. Fermat pointed out that *any* equation in two variables defines a curve. On top of that, if a curve is described algebraically, one naturally wants to find its tangent algebraically, as well. In *La Géométrie*, Descartes described a complicated algebraic method for finding tangents. He used the parabola as his example, in part because that allowed him to demonstrate that his method found the right answer, i.e., the one discovered by Apollonius.

Fermat had a slightly better method for tangents. Moreover, he realized that there was a connection between the problem of finding a tangent and the problem of finding maximum and minimum values. In both cases, Fermat's key insight was to connect the problem to what we would call "double roots." If you imagine a line cutting a curve in two nearby points and translate that into algebra, you get an equation with two roots close together. As the points converge, the line becomes closer and closer to a tangent. So a tangent occurs when the two roots become equal — i.e., when the equation has a double root. This was useful because there were known criteria to detect a double root.

Fermat soon developed a more effective method for doing the same thing: he wrote the x-coordinates of the two intersection points as a and $a + e$, so that the corresponding y-coordinates are a^2 and $a^2 + 2ae + e^2$. The slope of the line connecting them is

$$\frac{(a^2 + 2ae + e^2) - a^2}{(a + e) - a} = \frac{2ae + e^2}{e} = 2a + e.$$

To make the two points coincide, set $e = 0$ and you get the slope.

All of these mathematicians were essentially "doing calculus," but of course there wasn't any "calculus" yet. We might call this the *heroic period* of calculus: individual mathematicians solving tangent and maximum problems for individual curves, sometimes with specially created methods. When supposedly general methods were proposed, they tended to work well only for polynomials. The cycloid, which is not a polynomial curve, was interesting for that very reason. To do anything with it required tricks quite different from ones that worked for the parabola. If someone came up with a new curve, one would have to start from scratch, because there was no general method.

It is in this context that we can understand how exciting it would have been to read the title[1] of Leibniz's famous first article on calculus, published in 1688: "A new method for maxima and minima and the same for tangents, unhindered by either fractional or irrational powers, and a unique kind of calculus for that." What that title says is that Leibniz has discovered a way to compute tangents and to solve maximum and minimum problems that will work for *any* equation. Even more, he says he has developed "a calculus"— that is, a straightforward computational way to solve this kind of problem.

Reading Leibniz's paper is a remarkable experience. He says that we should take an arbitrary quantity and call it dx. If there is a formula relating y to x that defines a curve, then dy is defined to be whatever quantity makes dy/dx the slope of the tangent line.[2] And then he starts teaching the rules: $d(y + z) = dy + dz$, $d(yz) = y\,dz + z\,dy$, $d(x^n) = nx^{n-1}\,dx$, etc. "Just calculate," we can imagine him saying. "Finding dy lets you find dy/dx, and hence the tangent line. Setting $dy = 0$ finds maximum values of y. Here's an example." The article reads like a *Cliff's Notes* version of a calculus book. Leibniz gives only the slightest hint of what all this *means* and does not explain how he figured out these rules. He just gives the method: No need to think, just calculate this way and the answers will come out. For the parabola, Leibniz's instructions move you at once from $y = x^2$ to $dy = 2x\,dx$.

We can think of Leibniz's calculus as the beginning of a new period. It was no longer necessary to go to heroic extremes to solve this kind of problem. There was a method, and it just worked. Now it was a matter of exploring and extending the method. Because the method was fundamentally algebraic, just a bunch of rules for manipulating things, we might call this the *algebraic period* of the calculus.

---

[1] We quote the translation in [167], which includes the paper both in the original Latin and in English.

[2] Not in these exact words, but close.

Newton and Leibniz are considered joint inventors of the calculus because both of them invented methods to solve the same set of problems, but Newton's approach was more intuitive and more physical. He emphasized the idea of a rate of change, which he called the *fluxion* of a variable ("flowing") quantity. For Newton, x is a quantity that changes over time, and $\dot{x}$ is simply its rate of change, which he doesn't really define. Instead, he would say things like "when a moment of time o passes, x becomes $x + \dot{x}o$." So a Newtonian computation of the slope of the tangent to the parabola would look sort of like this. The parabola is given by $y = x^2$. When a moment of time o passes, y becomes $y + \dot{y}o$ and x becomes $x + \dot{x}o$. So we will have

$$y + \dot{y}o = (x + \dot{x}o)^2 = x^2 + 2x\dot{x}o + \dot{x}^2 o^2.$$

Subtracting $y = x^2$ and simplifying we get

$$\frac{\dot{y}}{\dot{x}} = \frac{2xo + \dot{x}o^2}{o} = 2x + \dot{x}o$$

now "let the augments vanish" to get the "ultimate ratio," $\dot{y}/\dot{x} = 2x$.

Leibniz's version was the more successful one, as we can see from the fact that his reference to "a calculus" has become the name of the whole subject. It quickly emerged that one should understand Leibniz's dx as an "infinitely small" quantity. Jakob and Johann Bernoulli developed and taught this point of view, which was then included in the first calculus textbook, by the Marquis de l'Hospital, published in 1696.

L'Hospital was a competent French mathematician. Moreover, being a Marquis, he had lots of money. Johann Bernoulli needed a job, so l'Hospital hired him as his calculus tutor. It was from Bernoulli's notes that l'Hospital put together his book, whose title translates to "Analysis of the Infinitely Small for the Understanding of Curved Lines." The book treats only the differential calculus, and it is indeed focused on using the calculus to understand curves. Finding tangents is the first and easiest part of this "understanding of curves."

As the title indicates, l'Hospital's book is all about infinitely small quantities such as dx and dy. But how do those work? The idea is roughly like this: if there is a normal (finite nonzero) number around, an infinitely small change does not matter. So, for example, $x^2 + 2dx$ is the same as x^2 for any such number x. But when there is no finite number around, then infinitely small things do matter: $2dx$ is not 0. It gets even more complicated when you run into products of infinitely small things: such a product is supposed to be "infinitely smaller": it doesn't matter in comparison to things that are merely infinitely small. So here's how we prove that Leibniz's rule is correct in the case of $y = x^2$. When x becomes $x + dx$, y becomes $y + dy$, so

$$y + dy = (x + dx)^2 = x^2 + 2x\,dx + (dx)^2 = x^2 + 2x\,dx.$$

That last equal sign is the hard one; it is there because $(dx)^2$ is infinitely smaller than $2x\,dx$, so it can be thrown out. At least so says l'Hospital.

This is recognizably the same computation as Newton's, though the underlying ideas are different: In one case, there are rates of change; in the other, everything is static but we use infinitely small increments. Both methods had their strange parts. In Newton's approach, there was this "moment of time" o that was nonzero until we "let it vanish." In Leibniz there were the "infinitely small things don't matter except when they do." This troubled some people, but most mathematicians seemed to learn to accept it. Their fundamental argument was a practical one: it worked. There were criticisms, notably from George Berkeley (see page 47), but they did not have much impact. Mathematicians took the calculus and ran with it. The main contributor in the 18th century was Leonhard Euler. He realized that Newton's laws of motion could be expressed using Leibniz's differentials, so one could do physics using them. The calculus became the key to everything. Physical problems could be reduced to equations involving differentials, and solving these *differential equations* allowed one to predict what would happen. Geometric questions could be analyzed in a similar way. It all boiled down to using the calculus intelligently, and there has never been a calculator more talented than Euler.

The one time Euler tried to explain what he was doing was when he wrote his own introduction to the differential calculus. The result is very strange: He agrees that a number smaller than any positive number must be zero, so infinitesimals such as dx *are zero*, so dy/dx really is $0/0$. But he argues that not all $0/0$ expressions are alike, and that if one considers where the zeros had come from one could assign a value to them. The idea seems to be that 0 is not just a number, it is a number with a memory! So the 0 that appears in the numerator and denominator of $(x^2 - 4)/(x - 2)$ when we set $x = 2$ "remembers" enough information to let us know that in this case $0/0 = 4$.

Why worry about foundations when there were so many new problems to solve? The answer turned out to be "in order to explain it better." After the French Revolution, mathematicians were asked to teach the calculus not only to other gifted mathematicians, but also to future engineers and civil servants. That required coming up with clear explanations, which is hard to do if the underlying concepts are not clear in the first place. Hence, beginning with Lagrange and Cauchy and continuing until late in the 19th century, mathematicians worked on laying the foundations for the calculus. They looked for clear definitions that did not depend on physical intuition about moving or flowing things or on philosophical ideas about infinitely small quantities.

Lagrange was the one who invented the "derivative function" and the notation $f'(x)$. His goal was to do Leibniz one better and reduce the whole of calculus to algebra. Here's how he wanted to do it. Given a function $f(x)$, use algebra to find a formula

$$f(x + h) = f(x) + Ah + Bh^2 + \ldots$$

Then we define $f'(x) = A$. The example of the parabola becomes very simple: Since $(x + h)^2 = x^2 + 2xh + h^2$, the derivative of $f(x) = x^2$ is $f'(x) = 2x$. The problem was that the method is much harder to use for more complicated functions, such as the sine.

It was in textbooks written by Cauchy that the notion of a limit was finally introduced and used to give a *definition* of the derivative:

$$f'(x) = \lim_{h \to 0} \frac{f(x + h) - f(x)}{h}.$$

Cauchy's notion of what "limit" meant was still not quite precise, however, and it took a few more decades for everything to get sorted out. Nonetheless, it all *did* get sorted out. While mathematicians found examples where the easy assumptions of Euler's time would fail, they also discovered precise conditions of validity and showed that those conditions usually held in practical situations. The new definition made things clearer, but did not invalidate any of the earlier work. This is why Judith Grabiner says[3] that

> The derivative was first *used*; it was then *discovered*; it was then *explored and developed*; and it was finally *defined*.

Derivatives are still fundamental to the understanding of curved lines, of surfaces, and also of more complicated geometrical objects; this is the subject of differential geometry. They are one of the basic tools used to describe how things change and evolve, mostly via differential equations. And while the foundational issues are resolved, there are still plenty of problems for today's heroes to solve.

For a Closer Look: A famous and readable account of the story of the derivative is [70]. Historians have done a lot of work on the beginnings of calculus. The results are well summarized in standard references such as [99] and [76], and [75].

---

[3]In the introduction to [70].

What to Read Next

The history of mathematics is a huge and fascinating subject. Since this book can hardly do more than pull back the curtain and give its readers a peek, it is important also to provide a guide to the literature.

Selective bibliographies such as this one are always the result of the authors' personal preferences, and they involve making judgments that are necessarily subjective and often matters of degree. We have used two main criteria in selecting and discussing books. First, we have selected books that are not too hard to read and do not have too many mathematical or historical prerequisites. Second, we have tried to choose books that are reliable sources. Of course, the history of mathematics is history, not mathematics. As in all history writing, there is often room for disagreement among historians; sometimes such disagreements even make it more interesting to learn the subject. Still, we have mostly avoided books that are widely considered to be out of date, highly speculative, or too prone to error.

What we give here is a small selection of books, ones that we think are especially noteworthy. See the notes at the end of each Sketch and at the end of each section of the Overview for many other references. Throughout, we refer to books by their title followed by a number in brackets. The numbers refer to the more complete citations in the bibliography.

The Reference Shelf

Let's first consider what might be present on an ideal shelf of reference books. These are books to dip into for answers to specific questions or for an overview of a particular period.

The first book that should be on our reference shelf is a big, formal history of mathematics. There are quite a few such books. Some of them were written as textbooks for college courses (it's easy to tell — they contain exercises!), others are aimed at a more general readership, and a few are really aimed at professional mathematicians. Of the ones available today, the best is probably *A History of Mathematics*, by Victor J. Katz [99]. It's a huge book with lots of information in it. Katz is aware of current historical research and provides good references. His book is not the easiest to read of the big histories, but it's the first one we would look at.

Several other one-volume histories have their virtues. Howard Eves's *An Introduction to the History of Mathematics* [53] and David M. Burton's *The History of Mathematics* [20] are both intended as college textbooks, and both are more accessible and less hefty than Katz, but both are showing their age. Ivor Grattan-Guinness's *The Rainbow of Mathematics* [76], which is intended for a broader audience but is still somewhat technical, is interesting because it pays much more attention than the others to recent mathematics and to applied mathematics. Open Katz, Eves, or Burton in the middle and you'll probably find a chapter on medieval mathematics; open *Rainbow* in the middle and you'll find yourself in the eighteenth century. This accurately reflects the huge explosion in the quantity of mathematics produced since early modern times. On the other hand, it means that many of the topics discussed by Grattan-Guinness are more advanced and therefore make more mathematical demands on the reader. Roger Cooke's *The History of Mathematics: A Brief Course* [30] is based on the course the author taught at the University of Vermont. As he says in the introduction, the book reflects his interests and is sometimes idiosyncratic, but it is interesting and fun to read. Luke Hodgkins' *A History of Mathematics* is shorter than Katz or Cooke and very much on top of recent trends. Jacqueline Stedall's *Mathematics Emerging* is intended as a new style of textbook based on direct inspection of historical texts.

A good library's reference shelf should contain the *Companion Encyclopedia of the History and Philosophy of the Mathematical Sciences* [74], a two-volume set edited by Grattan-Guinness. Like most books that collect articles from many authors, this is uneven, but it is still a great source for short histories of specific subjects and for bibliographical references. Unfortunately, it is now out of print.

Many of the books we mention are discussed in a reference book called *Landmark Writings in Western Mathematics 1640–1940* [73], also edited by Grattan-Guinness. Each article discusses an important book in detail, tracing editions and content. Also worth looking for in the library is the *Dictionary of Scientific Biography* [67], which exists both as a multi-volume set and as an online resource. It contains short biographies of scientists and mathematicians. The biographies in the *DSB* are often the best starting point for serious study of individual mathematicians.

Selin and D'Ambrosio's *Mathematics Across Cultures* [157] is also a collection of articles, in this case focusing on non-Western cultures, including some ancient cultures. Since this topic is sometimes under-represented in the big histories, this book is a valuable complement. Non-European roots of mathematics are also examined in George Gheverghese Joseph's *The Crest of the Peacock* [94].

Two other collections of articles are also good sources. The first is *From Five Fingers to Infinity* [170], edited by Frank Swetz. It collects articles on the history of mathematics extracted from journals aimed at a broad mathematical audience (such as *Mathematics Teacher* and *Mathematics Magazine*). Some of these articles are true gems, and the book as a whole is useful and fun to read. Unfortunately, it is now out of print, but sections of it are being reprinted by Dover as independent books. So far, there are [172], on the early modern period, and [171], on the 19th and 20th centuries. Similar, but with a much broader range, is *The World of Mathematics* [133], a four-volume collection put together by James R. Newman that tries to give nonspecialists access to the world of mathematics. It includes fiction, history, biography, expository articles, and lots more. The specifically historical articles are in the first volume.

A few more specialized books are worth noting. *The Historical Roots of Elementary Mathematics* [19], by Lucas N. H. Bunt, Phillip S. Jones, and Jack D. Bediant, concentrates only on the elementary parts of mathematics and does a good job of explaining their history. Florian Cajori's *A History of Mathematical Notations* [22] is a reference book on how mathematical notation developed, and it can often give the right answer to the "who was the first to use this symbol" questions (but see the "History Online" section below for a modern rival). A more readable account of symbolism is Joseph Mazur's *Enlightening Symbols*[122]. For a scholarly examination of a wide range of numeration systems, see Stephen Chrisomalis, *Numerical Notation: A Comparative History* [29].

The role of women in mathematical history is explored in several good books, several of which have virtually identical titles. Lynn Osen's *Women in Mathematics* [136] focuses on the lives of eight female mathematicians spread over the time spectrum from the Greek era to the early 20th century. Despite some discrepancies with more recent historical research, the stories she tells are engaging and informative. The introductory and concluding sections are thought-provoking expositions of the circumstances that made women in pre-20th century mathematical history almost invisible. *Women of Mathematics* [81], which calls itself "a biobibliographical sourcebook," is a compilation of brief biographies and literature references for 43 female mathematicians, all but three of whom lived during the 19th or the 20th century. The brief biographies are quite readable and make good starting places for student research on particular women. *Notable Women in Mathematics* [127] is quite similar, and perhaps a little easier to read. *Pioneering Women in American Mathematics*, by Judy Green and Jeanne LaDuke, focuses on American women who obtained Ph.D.s in mathematics before 1940. Somewhat more sophisticated is Claudia Henrion's *Women in Mathematics* [87], which uses the stories of nine contemporary female mathematicians (most of whom are still active) to explore the broader professional context for women in the field. *Complexities: Women in Mathematics* is almost a handbook for women in mathematics, containing biographies, analytical essays, research findings, and even a few expository papers.

Finally, some books are just fun. Howard Eves has put together several collections of anecdotes about mathematicians in his *Mathematical Circles* [55, 57, 56] series. These are pleasant to read and are a good source of stories that can add some interest to a class. *Mathematical Apocrypha* [109] and *Mathematical Apocrypha Redux* [110], by Steven Krantz, are similar. Books of quotes can also be fun. Two good ones are *Memorabilia Mathematica* [126] (an older book) and *Out of the Mouths of Mathematicians* [152] (a newer one). You can also find several collections of mathematical quotations on the Internet. Finally, there is *Stamping Through Mathematics* [177], by Robin Wilson. This beautiful book reproduces stamps featuring mathematics and mathematicians, organized in historical order. The result is fascinating and a potential source of interesting visuals for a class presentation.

Twelve Historical Books You Ought to Read

The books we have described so far are useful as reference books, but few readers will actually want to read one from cover to cover. In most cases, that's not what they are designed for. In this section, we give a short list of history books that we think are both readable and worth reading. Some of these books were written by historians, others by writers who relied on the historical research of others. Except as indicated, we feel that these books are reliable sources; still, every reader of historical writing should "trust, but verify."

The History of Mathematics: A Very Short Introduction is Jacqueline Stedall's introduction to the goals and methods of historians, as well as to the history of mathematics. This little book is an excellent way to get your bearings before tackling modern historical texts. It's short and inexpensive, so start here.

Tobias Dantzig's *Number: the Language of Science* [36] is an insightful, eloquent chronicle of the evolution of the number concept from its primitive beginnings in prehistory to the modern sophistication of complex and transfinite numbers. Along the way, Dantzig's story touches on many topics in early algebra and geometry, as well, providing a finely crafted, unifying perspective on mathematical history.

William Dunham's *Journey Through Genius: The Great Theorems of Mathematics* [48] surveys the history of mathematics by focusing on a small selection of important theorems. Each chapter contains an extensive historical introduction to the subject in question, an account of the proof of the theorem, and then a summary of what happened to the result after it was first proved. The result is a book with real mathematical content that is definitely still readable.

Books on ancient mathematics are often quite difficult, which is a pity because the subject is fascinating. *The Archimedes Codex* [132], by Reviel Netz and William Noel, is an exception. It tells the story of the "Archimedes Palympsest," a manuscript of Archimedes discovered in the late 19th century, then lost, then rediscovered and studied in the early 21st century. The story is exciting, and along the way we learn a lot about ancient mathematics and modern research.

Medieval mathematics is often passed by, but those too were interesting times. Nancy Marie Brown's *The Abacus and the Cross* [18] tells the amazing story of Gerbert d'Aurillac, who later became Pope Sylvester II. Gerbert was one of the pioneers of European mathematics, going to Spain to learn Arabic mathematics and bringing the knowledge back to schools in France and the Holy Roman Empire. Brown may overstate the historical importance of Gerbert, but she tells a good story.

Glen van Brummelen's *Heavenly Mathematics* is a historical introduction to spherical trigonometry that shows that it's possible for trigonometry to be fascinating. One can read it to learn how to solve such problems as "when will the sun rise tomorrow," but we suggest you read it for the story.

Men of Mathematics [12], by E. T. Bell, is a collection of readable and engaging biographies of mathematicians (not all male!) throughout history. Bell knows how to write an entertaining story, and he pulls out all the stops to get the reader to really care about the lives of the mathematicians he profiles. The book has lost some of its original popularity, not (or at least not primarily) because of the politically incorrect title, but rather because Bell takes too many liberties with his sources. (Some critics would say "because he makes things up.") The book is fun to read, but don't rely solely on Bell for facts.

Dava Sobel's *Longitude* [166] provides a good picture of the interactions among mathematics, astronomy, and navigation in the 18th century as it chronicles the life of John Harrison, the master clockmaker who solved the problem of reliable timekeeping at sea.

Robert Osserman's *Poetry of the Universe* [137] is a short and readable account of how ideas about geometry have affected our view of the universe in which we live. Along the way, Osserman includes lots of material about the history of geometry, including a good discussion of non-Euclidean geometries.

Though it's not exactly a history book, Barry Mazur's *Imagining Numbers (Particularly the Square Root of Minus Fifteen)* [121] tells a good part of the story of complex numbers. Mazur's book is directed at "people who have had *no* background in math, who never paid attention to the subject, or who abhorred it, in high school, but who are happy to spend hours thinking over a phrase of poetry." The book invites them to use their talents to imagine numbers such as the square root of -15.

Since the mathematics of the 20th century is very technical, it is hard to get a handle on its history. Biographies and interviews are one way to get a feeling for what was going on. Two books that use biographies as a point of entry to the story are David Salsburg's *The Lady Tasting Tea* [151], a history of statistics in the 20th century, and Benjamin Yandell's *The Honors Class* [178], which focuses on the lives of the mathematicians who worked on the problems proposed by David Hilbert at the International Congress of Mathematicians in 1900.

History Online

Historical information is not found just in books. These days, you can also find it on the Internet. There are lots of websites dealing with the history of mathematics. As usual, the biggest problem is reliability. Since it's so easy to create a website, one is not always certain about the quality of the information. Here, too, it's best to "trust, but verify."

This section points out a few of the more interesting things that are out there, without trying to be too complete or detailed. We list only the sites that seem particularly useful to us. We don't give web addresses, since they change so much, but we give enough information to make it easy to find the site with your favorite search engine.

One must start, of course, with *Wikipedia*. It can be surprisingly good on most things related to mathematics and its history. The key, of course, is to check the sources and follow the links given at the end.

The *MacTutor History of Mathematics Archive* is well known for its huge collection of brief biographies of mathematicians. The biographical accounts include quotes, photographs (when possible), and often other material too.

The Internet is an excellent way to get your hands on original source material. Many older books have been scanned and made available at such sites as *Gallica, Europeana,* and *Early English Books Online.* Euclid's *Elements* is online in both Greek and English. All of Euler can be found at *The Euler Archive.* Most scholarly journals have put their complete archives online. Whether you want to read the *Acta Eruditorum* or the *Annals of Mathematics,* the first step is to try to find it online. (But yes, sometimes you have to pay.)

Two websites, both maintained by Jeff Miller, a teacher at Gulf High School in New Port Richey, Florida, are useful and fun. The first is *Earliest Uses of Various Mathematical Symbols.* It deals with the history of mathematical notation and other symbols. In a way, it's the modern-day reply to Florian Cajori's book mentioned above. Its sister site, *Earliest Known Uses of Some of the Words of Mathematics,* deals with mathematical terms and their origin. Both sites contain lots of material that can be used to enrich a mathematics class.

There is a history of mathematics subpage at the *Math Forum @ Drexel University,* an enormous and very useful website hosted by Drexel University that collects mathematical resources of all kinds. The *Forum* maintains a large collection of history of mathematics links (click on "Math Topics" then choose "History/Biography").

Finally, we should mention the online journal *Convergence,* a publication of the Mathematical Association of America. This focuses specifically on the history of mathematics and its use in the classroom. It offers lots of resources — articles, visual resources called "mathematical treasures," quotations, even an "on this day" feature.

When They Lived

Abu Kamil (ca. 850 – ca. 930)
Abel, Niels Henrik (1802 – 1829)
Agnesi, Maria Gaetana (1718 – 1799)
Aiken, Howard (1900 – 1973)
Alberti, Leone Battista (1404 – 1472)
Alexander the Great (356 – 323 B.C.)
Abu-l'Wafa' (940 – 998)
al-Bīrūnī, Abu Rayhan (973 – 1052)
al-Karajī, Abū Bakr (fl. ca. 980 – 1030)
al-Kāshī, Jamshīd (ca. 1380 – 1429)
al-Khāyammī, 'Umar (1048 – 1131)
al-Khwārizmī, Muḥammad ibn Mūsa (ca. 780 – ca. 850)
Apollonius of Perga (fl. ca. 200 B.C.)
Aquinas, Saint Thomas (ca. 1225 – 1274)
Archimedes (ca. 285 – 212 B.C.)
Argand, J. R. (1768 – 1822)
Aristaeus (the Elder) (fl. ca. 350 – 330 B.C.)
Aristotle (384 – 322 B.C.)
Arnauld, Antoine (1612 – 1694)
Artin, Emil (1898 – 1962)
Āryabhaṭa (476 – 550)
Atanasoff, John (1903 – 1995)

Babbage, Charles (1791 – 1871)
bar Hiyya, Abraham (1065 – 1136)
Berkeley, George (1685 – 1753)
Bernoulli, Daniel (1700 – 1782)
Bernoulli, Jakob (1655 – 1705)
Bernoulli, Johann (1667 – 1748)
Berry, Clifford (1918 – 1963)
Bhāskara I (ca. 600 – ca. 680)
Bhāskara II (1114 – 1185)
Bolyai, János (1802 – 1860)
Bombelli, Rafael (1526 – 1572)
Boole, George (1815 – 1864)
Bourbaki, Nicolas (fl. 1935 – 1955)
Brahe, Tycho (1546 – 1601)
Brahmagupta (598 – 665)
Briggs, Henry (1561 – 1630)
Brunelleschi, Filippo (1377 – 1446)
Bürgi, Joost (1552 – 1632)

Cantor, Georg (1845 – 1918)
Cardano, Girolamo (1501 – 1576)
Cauchy, Augustin Louis (1789 – 1857)
Cavalieri, Bonaventura (1598 – 1647)
Châtelet, Emilie de (1706 – 1749)
Chudnovsky, David (b. 1952)
Chudnovsky, Gregory (b. 1947)
Chuquet, Nicholas (1445 – 1488)
Copernicus, Nicolaus (1473 – 1543)

d'Alembert, Jean le Rond (1717 – 1783)
Darwin, Charles (1809 – 1882)
da Vinci, Leonardo (1452 – 1519)
de Colmar, Charles (1785 – 1870)
Dedekind, Richard (1831 – 1916)
del Ferro, Scipione (1465 – 1526)
della Francesca, Piero (ca. 1415 – 1492)
De Moivre, Abraham (1667 – 1754)
De Morgan, Augustus (1806 – 1871)
Desargues, Girard (1591 – 1661)
Descartes, René (1596 – 1650)
Diophantus (fl. ca. 250?)
Dirichlet, Lejeune (1805 – 1859)
Dürer, Albrecht (1471 – 1528)

Eckert, J. Presper (1919 – 1995)
Edgeworth, Francis (1845 – 1926)
Einstein, Albert (1879 – 1955)
Euclid (fl. 3rd century B.C.)
Eudemus (fl. 4th century B.C.)
Eudoxus (fl. 4th century B.C.)
Euler, Leonhard (1707 – 1783)

Fermat, Pierre de (1601 – 1665)
Ferrari, Lodovico (1522 – 1565)
Fibonacci (Leonardo of Pisa) (ca. 1170 – ca. 1250)
Fincke, Thomas (1561 – 1656)
Fiore, Antonio Maria (fl. 1530)
Fisher, R. A. (1890 – 1962)
Flowers, Tommy (1905 – 1998)
Fontana, Niccolò (Tartaglia) (1499 – 1557)
Fourier, Joseph (1768 – 1830)

Galileo Galilei (1564 – 1642)
Galois, Évariste (1811 – 1832)
Galton, Sir Francis (1822 – 1911)
Gardiner, William (fl. 1740s)

Garfield, President James (1831 – 1881)
Gauss, Carl Friedrich (1777 – 1855)
Gelfond, Alexsandr (1906 – 1968)
Gerbert d'Aurillac (Pope Sylvester II) (945 – 1003)
Germain, Sophie (1776 – 1831)
Girard, Albert (1595 – 1632)
Gosset, William S. (1876 – 1937)
Graunt, John (1620 – 1674)
Gregory, James (1638 – 1675)

Hadamard, Jacques (1865 – 1963)
Halley, Edmund (1656 – 1742)
Hamilton, William Rowan (1805 – 1865)
Harriot, Thomas (ca. 1560 – 1621)
Hérigone, Pierre (d. ca. 1643)
Hermite, Charles (1822 – 1901)
Heron (fl. 60 A.D.)
Hilbert, David (1862 – 1943)
Hipparchus (fl. 2nd century B.C.)
Hippocrates of Chios (fl. 5th century B.C.)
Hollerith, Herman (1860 – 1929)
Huygens, Christiaan (1629 – 1695)
Hypatia (ca. 370 – 415)

Jacquard, Joseph-Marie (1752 – 1834)
John of Seville (fl. 1135 – 1153)
Jones, William (1675 – 1749)

Kanada, Yasumasa (b. 1949)
Kepler, Johannes (1571 – 1630)
Khayyam, Omar (1048 – 1131)
Klein, Felix (1849 – 1925)
Kronecker, Leopold (1823 – 1891)
Kummer, Ernst (1810 – 1893)

Lagrange, Joseph-Louis (1736 – 1813)
Lambert, Johann (1728 – 1777)
Lamé, Gabriel (1795 – 1870)
Laplace, Pierre Simon (1749 – 1827)
Legendre, Adrien-Marie (1752 – 1833)
Leibniz, Gottfried Wilhelm (1646 – 1716)
Leonardo of Pisa (Fibonacci) (ca. 1170 – ca. 1250)
L'Hospital, Marquis de (1661 – 1704)
Lindemann, Ferdinand (1852 – 1939)
Liouville, Joseph (1809 – 1882)
Liu Hui (fl. 220 – 265)

Lobachevsky, Nicolai (1792 – 1856)
Lovelace, Augusta Ada (1815 – 1852)

Mahāvīra (fl. 850)
Mauchly, John (1907 – 1980)
Menaechmus (fl. 4th century B.C.)
Méré, Chevalier de (1607 – 1684)
Mersenne, Marin (1588 – 1648)
Müller, Johannes (Regiomontanus) (1436 – ca. 1476)

Napier, John (1550 – 1617)
Newman, Max (1897 – 1984)
Newton, Isaac (1642 – 1727)
Noether, Emmy (1882 – 1935)
Nunes, Pedro (1502 – 1578)

Oresme, Nicole (ca. 1320 – 1382)
Otho, Lucius Valentin (ca. 1550 – 1603)
Oughtred, William (1575 – 1660)

Pacioli, Luca (1445 – 1517)
Pandrosian (fl. 284 – 305)
Pappus (fl. 300 – 350)
Pascal, Blaise (1623 – 1662)
Peano, Giuseppe (1858 – 1932)
Pearson, Karl (1857 – 1936)
Peirce, Benjamin (1809 – 1880)
Peirce, Charles Sanders (1839 – 1914)
Petty, William (1623 – 1687)
Pitiscus, Bartholomew (1561 – 1613)
Plato (427 – 347 B.C.)
Playfair, John (1748 – 1819)
Poincaré, Henri (1854 – 1912)
Poncelet, Jean Victor (1788 – 1867)
Proclus (410? – 485)
Ptolemy, Claudius (ca. 100 – ca. 170)
Pythagoras (of Samos) (fl. 6th ca. B.C.)

Quetelet, Lambert (1796 – 1874)

Rahn, Johann (1622 – 1676)
Recorde, Robert (ca. 1510 – 1558)
Regiomontanus (Johannes Müller) (1436 – ca. 1476)
Rheticus, Georg Joachim (1514 – 1574)
Ribet, Kenneth (b. 1948)
Riemann, Bernhard (1826 – 1866)
Robert of Chester (fl. 1140 – 1150)

Roberval, Gilles de (1602 – 1675)
Rudolff, Christoff (1499 – 1543)
Russell, Bertrand (1872 – 1970)

Saccheri, Girolamo (1667 – 1733)
Saint Vincent, Grégoire de (1584 – 1667)
Schneider, Theodor (1911 – 1988)
Shanks, William (1812 – 1882)
Shannon, Claude (1916 – 2001)
Spinoza, Baruch (1632 – 1677)
Stevin, Simon (1548 – ca. 1620)
Stifel, Michael (ca. 1487 – 1567)
Sylvester, James Joseph (1814 – 1897)

Tartaglia (Niccolò Fontana) (1499 – 1557)
Taylor, Richard (b. 1962)
Thābit ibn Qurra (836 – 901)
Thales (fl. 6th century B.C.)
Theon (fl. 4th century A.D.)
Tukey, John (1915 – 2000)
Turing, Alan (1912 – 1954)

van Schooten, Frans (ca. 1615 – 1660)
Viète, François (1540 – 1603)
von Neumann, John (1903 – 1957)

Wallis, John (1616 – 1703)
Weierstrass, Karl (1815 – 1897)
Widman, Johann (ca. 1462 – ca. 1500)
Wiles, Andrew (b. 1953)
Witelo, Erazmus (fl. 1270 – 1280)
Wolfskehl, Paul (1856 – 1906)

Yule, G. Udny (1871 – 1951)

Zu Chongzhi (429 – 501)
Zuse, Konrad (1910 – 1995)

Bibliography

[1] Irving Adler. *Probability and Statistics for Everyman*. The John Day Co., New York, 1963.

[2] Muhammad ibn Musa al Khuwarizmi. *The Algebra of Mohammed ben Musa, edited and translated by Frederic Rosen*, volume 1 of *Islamic Mathematics and Astronomy*. Institute for the History of Science at the Johann Wolfgang Goethe University, Frankfort am Main, 1997. Bilingual reprint of the original edition, London 1830–31.

[3] Donald J. Albers and Gerald L. Alexanderson, editors. *More Mathematical People*. Birkhäuser, Boston, 1990.

[4] Donald J. Albers and Gerald L. Alexanderson, editors. *Mathematical People*. A K Peters Ltd., Wellesley, MA, second edition, 2008.

[5] Donald J. Albers and Gerald L. Alexanderson, editors. *Fascinating Mathematical People*. Princeton University Press, Princeton, NJ, 2011.

[6] Kirsti Andersen. *The Geometry of an Art: The History of the Mathematical Theory of Perspective from Alberti to Monge*. Springer, New York, 2007.

[7] Benno Artmann. *Euclid: The Creation of Mathematics*. Springer-Verlag, Berlin, Heidelberg, New York, 1999.

[8] Marcia Ascher. *Ethnomathematics: A Multicultural View of Mathematical Ideas*. Brooks/Cole, Pacific Grove, CA, 1991.

[9] William Aspray, editor. *Computing Before Computers*. Iowa State University Press, Ames, IA, 1990.

[10] J. K. Baumgart, D. E. Deal, B. R. Vogeli, and A. E. Hallerberg, editors. *NCTM Thirty-first Yearbook: Historical Topics for the Mathematics Classroom*. National Council of Teachers of Mathematics, Washington, DC, 1969, revised 1989.

[11] Petr Beckmann. *A History of Pi*. Barnes & Noble, New York, 1993.

[12] E. T. Bell. *Men of Mathematics*. Simon & Schuster, New York, 1937.

[13] J. L. Berggren. *Episodes in the Mathematics of Medieval Islam*. Springer-Verlag, Berlin, Heidelberg, New York, 1986.

[14] Lennart Berggren, Jonathan Borwein, and Peter Borwein. *Pi: A Source Book*. Springer-Verlag, Berlin, Heidelberg, New York, 1997.

[15] William P. Berlinghoff and Fernando Q. Gouvêa. *Pathways from the Past I: Using History to Teach Numbers, Numerals, and Arithmetic*. Oxton House, Farmington, ME, 2010.

[16] William P. Berlinghoff and Fernando Q. Gouvêa. *Pathways from the Past II: Using History to Teach Algebra*. Oxton House, Farmington, ME, 2013.

[17] Carl B. Boyer. *History of Analytic Geometry*. Scripta Mathematica, New York, 1956.

[18] Nancy Marie Brown. *The Abacus and the Cross: The Story of the Pope Who Brought the Light of Science to the Dark Ages*. Basic Books, New York, 2010.

[19] Lucas N. H. Bunt, Phillip S. Jones, and Jack D. Bediant. *The Historical Roots of Elementary Mathematics*. Dover Publications, New York, 1976.

[20] David M. Burton. *The History of Mathematics*. McGraw-Hill, New York, fourth edition, 1998.

[21] Florian Cajori. *A History of Mathematics*. AMS Chelsea Publishing, Providence, RI, fifth edition, 1991.

[22] Florian Cajori. *A History of Mathematical Notations*. Dover Publications, New York, 1993.

[23] Ronald Calinger, editor. *Vita Mathematica: Historical Research and Integration with Teaching*. Mathematical Association of America, Washington, DC, 1996.

[24] Georg Cantor. *Contributions to the Founding of the Theory of Transfinite Numbers*. Dover Publications, New York, 1955.

[25] Girolamo Cardano. *Ars Magna, or the Rules of Algebra.* Dover Publications, New York, 1993.

[26] Girolamo Cardano. *Autobiography.* I Tatti Renaissance Library. Harvard University Press, Cambridge, MA, forthcoming. Edited by Thomas Cerbu and Anthony Grafton.

[27] Bettye Anne Case and Anne M. Leggett, editors. *Complexities: Women in Mathematics.* Princeton University Press, Princeton, NJ, 2005.

[28] Karine Chemla, editor. *The History of Mathematical Proof in Ancient Traditions.* Cambridge University Press, Cambridge and New York, 2012.

[29] Stephen Chrisomalis. *Numerical Notation: A Comparative History.* Cambridge University Press, Cambridge, 2010.

[30] Roger Cooke. *The History of Mathematics: A Brief Course.* John Wiley & Sons, New York, third edition, 2012.

[31] Julian Lowell Coolidge. *A History of the Conic Sections and Quadric Surfaces.* Dover, New York, 1968. Originally published by The Clarendon Press, Oxford, 1945.

[32] National Research Council. *The Mathematical Sciences in 2025.* National Academies Press, Washington, DC, 2013.

[33] Peter R. Cromwell. *Polyhedra.* Cambridge University Press, Cambridge, 1997.

[34] John N. Crossley and Alan S. Henry. Thus spake al-Khwārizmī: a translation of the text of Cambridge University Library ms. Ii.vi.5. *Historia Mathematica*, 17:103–131, 1990.

[35] S. Cuomo. *Ancient Mathematics.* Routledge, London and New York, 2001.

[36] Tobias Dantzig. *Number: The Language of Science.* Pi Press, New York, fourth edition, 2005. Original publication by Scribner, 1954.

[37] Lorraine Daston. *Classical Probability in the Enlightenment.* Princeton University Press, Princeton, NJ, 1988.

[38] Joseph Warren Dauben. *Georg Cantor: His Mathematics and Philosophy of the Infinite*. Princeton University Press, Princeton, NJ, 1990.

[39] Martin Davis. *The Universal Computer: The Road from Leibniz to Turing*. W. W. Norton, New York, 2000.

[40] R. Decker and S. Hirschfield. *The Analytical Engine*. Wadsworth, Belmont, CA, 1990.

[41] Richard Dedekind. *Essays in the Theory of Numbers*. Dover Publications, New York, 1963.

[42] René Descartes. *The Geometry of René Descartes: With a facsimile of the first edition*. Dover Publications, New York, 1954. Translated from the French and Latin by David Eugene Smith and Marcia L. Latham.

[43] Keith Devlin. *Mathematics: the New Golden Age*. Columbia University Press, New York, second edition, 1999.

[44] Keith Devlin. *The Unfinished Game: Pascal, Fermat, and the Seventeenth-Century Letter that Made the World Modern*. Basic Books, New York, 2008.

[45] Keith Devlin. *Leonardo and Steve: The Young Genius Who Beat Apple to Market by 800 Years*. Keith Devlin, 2011. Electronic book.

[46] Keith Devlin. *The Man of Numbers: Fibonacci's Arithmetic Revolution*. Walker and Co., New York, 2011.

[47] Yvonne Dold-Samplonius, Joseph W. Dauben, Menso Folkerts, and Benno van Dalen, editors. *From China to Paris: 2000 Years Transmission of Mathematical Ideas*, Stuttgart, 2002. Franz Steiner Verlag Wiesbaden GmbH.

[48] William Dunham. *Journey Through Genius: The Great Theorems of Mathematics*. John Wiley & Sons, New York, 1990.

[49] Euclid. *The Thirteen Books of Euclid's Elements*. Dover Publications, New York, 1956. Translated with introduction and commentary by Thomas L. Heath.

[50] Euclid. *Euclid's Elements: All Thirteen Books Complete in One Volume*. Green Lion Press, Santa Fe, NM, 2002. The Thomas L. Heath translation, Edited by Dana Densmore.

[51] Leonhard Euler. *Elements of Algebra.* Springer-Verlag, Berlin, Heidelberg, New York, 1984.

[52] Howard Eves. *A Survey of Geometry.* Allyn and Bacon, Boston, 1963.

[53] Howard Eves. *An Introduction to the History of Mathematics.* Holt, Rinehart and Winston, New York, fourth edition, 1976.

[54] Howard Eves. *Great Moments in Mathematics (after 1650).* Mathematical Association of America, Washington, DC, 1981.

[55] Howard Eves. *In Mathematical Circles.* Mathematical Association of America, Washington, DC, 2002. Originally published by Prindle, Weber & Schmidt, 1969.

[56] Howard Eves. *Mathematical Circles Adieu and Return to Mathematical Circles.* Mathematical Association of America, Washington, DC, 2003. Originally published as two separate volumes by Prindle, Weber & Schmidt, 1977 and 1987.

[57] Howard Eves. *Mathematical Circles Revisited and Mathematical Circles Squared.* Mathematical Association of America, Washington, DC, 2003. Originally published as two separate volumes by Prindle, Weber & Schmidt, 1972.

[58] Howard Eves and Carroll V. Newsom. *An Introduction to the Foundations and Fundamental Concepts of Mathematics.* Holt, Rinehart and Winston, New York, revised edition, 1965.

[59] John Fauvel and Jeremy Gray, editors. *The History of Mathematics, a Reader.* Macmillan Press Ltd., Basingstoke, 1988. Distributed in the U.S. by the Mathematical Association of America.

[60] John Fauvel and Jan van Maanen, editors. *History in Mathematics Education: an ICMI Study.* Kluwer Academic, Dordrecht, Boston, London, 2000.

[61] J. V. Field. *The Invention of Infinity: Mathematics and Art in the Renaissance.* Oxford University Press, Oxford and New York, 1997.

[62] David Fowler. 400 years of decimal fractions. *Mathematics Teaching,* 110:20–21, 1985. Published by the Association of Teachers of Mathematics, Lancashire, England.

[63] David Fowler. 400.25 years of decimal fractions. *Mathematics Teaching*, 111:30–31, 1985. Published by the Association of Teachers of Mathematics, Lancashire, England.

[64] David Fowler. *The Mathematics of Plato's Academy.* Oxford University Press/The Clarendon Press, Oxford and New York, second edition, 1999.

[65] Paulus Gerdes. *Geometry from Africa: Mathematical and Educational Explorations.* Mathematical Association of America, Washington, DC, 1999.

[66] Judith L. Gersting and Michael C. Gemignani. *The Computer: History, Workings, Uses & Limitations.* Ardsley House, New York, 1988.

[67] Charles Coulston Gillispie, editor. *Dictionary of Scientific Biography.* Scribner, New York, 1970–1980. 18 volumes.

[68] Herman H. Goldstine. *The Computer from Pascal to von Neumann.* Princeton University Press, Princeton, NJ, 1972.

[69] Enrique A. González-Velasco. *Journey through Mathematics: Creative Episodes in its History.* Springer, New York, 2011.

[70] Judith V. Grabiner. The Changing Concept of Change: The Derivative from Fermat to Weierstrass. *Mathematics Magazine*, 56(4):195–206, 1983. Reprinted in [170], [172], [72], and other places.

[71] Judith V. Grabiner. Why did Lagrange "prove" the Parallel Postulate? *American Mathematical Monthly*, 116(1):3–18, 2009. Reprinted in [72].

[72] Judith V. Grabiner. *A Historian Looks Back:* The Calculus as Algebra *and Selected Writings.* Mathematical Association of America, Washington, DC, 2010.

[73] I. Grattan-Guinness, editor. *Landmark Writings in Western Mathematics 1640–1940.* Elsevier B. V., Amsterdam, 2005.

[74] Ivor Grattan-Guinness, editor. *Companion Encyclopedia of the History and Philosophy of the Mathematical Sciences.* Routledge, London and New York, 1994.

[75] Ivor Grattan-Guinness, editor. *From the Calculus to Set Theory, 1630–1910: An Introductory History.* Princeton University Press, Princeton, NJ, 2000.

[76] Ivor Grattan-Guinness. *The Rainbow of Mathematics: A History of the Mathematical Sciences.* W. W. Norton, New York, 2000.

[77] Jeremy Gray. *Plato's Ghost: The Modernist Transformation of Mathematics.* Princeton University Press, Princeton, NJ, 2008.

[78] Jeremy Gray. *Worlds Out of Nothing: A Course in the History of Geometry in the 19th Century.* Springer, Berlin, Heidelberg, New York, 2010.

[79] Judy Green and Jeanne LaDuke. *Pioneering Women in American Mathematics: The pre-1940 PhD's.* American Mathematical Society, Providence, RI, 2009.

[80] Marvin Jay Greenberg. *Euclidean and non-Euclidean Geometries: Development and History.* W. H. Freeman and Company, New York, fourth edition, 2008.

[81] Louise S. Grinstein and Paul J. Campbell, editors. *Women of Mathematics.* Greenwood Press, Westport, CT, 1987.

[82] Ian Hacking. *The Emergence of Probability: A Philosophical Study of Early Ideas About Probability, Induction, and Statistical Inference.* Cambridge University Press, Cambridge, 1975.

[83] George Bruce Halsted. *Girolamo Saccheri's Euclides Vindicatus.* AMS Chelsea, Providence, RI, 1986. Originally published by Open Court, 1920.

[84] Julian Havil. *The Irrationals: A Story of the Numbers You Can't Count On.* Princeton University Press, Princeton, NJ, 2012.

[85] Cynthia Hay, editor. *Mathematics from Manuscript to Print. 1300–1600*, Oxford and New York, 1988. Oxford University Press/The Clarendon Press.

[86] J. L. Heilbron. *Geometry Civilized: History, Culture, and Technique.* Oxford University Press/The Clarendon Press, Oxford and New York, 1998.

[87] Claudia Henrion. *Women In Mathematics: The Addition of Difference.* Indiana University Press, 1997.

[88] C. C. Heyde and E. Seneta, editors. *Statisticians of the Centuries*. Springer-Verlag, Berlin, Heidelberg, New York, 2001.

[89] Victor E. Hill, IV. President Garfield and the Pythagorean Theorem. *Math Horizons*, pages 9–11, 15, February 2002.

[90] Jens Høyrup. *In Measure, Number, and Weight: Studies in Mathematics and Culture*. State University of New York Press, Albany, NY, 1994.

[91] Jens Høyrup. Subscientific mathematics: Observations on a premodern phenomenon. In *In Measure, Number, and Weight: Studies in Mathematics and Culture* [90], pages 23–43.

[92] Catherine Jami. *The Emperor's New Mathematics: Western Learning and Imperial Authority During the Kangxi Reign (1662–1722)*. Oxford University Press, Oxford and New York, 2012.

[93] Dick Jardine and Amy Shell-Gellasch, editors. *Mathematical Time Capsules: Historical Modules for the Mathematics Classroom*. Mathematical Association of America, Washington, DC, 2010.

[94] George Gheverghese Joseph. *The Crest of the Peacock: Non-European Roots of Mathematics*. Princeton University Press, Princeton, NJ, third edition, 2011.

[95] Robert Kaplan. *The Nothing That Is*. Oxford University Press, Oxford and New York, 2000.

[96] Victor Katz, editor. *The Mathematics of Egypt, Mesopotamia, China, India, and Islam: A Sourcebook*. Princeton University Press, Princeton, NJ, 2007.

[97] Victor Katz and Karen Dee Michalowicz, editors. *Historical Modules for the Teaching and Learning of Mathematics*. Mathematical Association of America, Washington, DC, 2005. Available as an e-book and on CD-ROM.

[98] Victor J. Katz, editor. *Using History to Teach Mathematics: An International Perspective*. Mathematical Association of America, Washington, DC, 2000.

[99] Victor J. Katz. *A History of Mathematics*. Addison-Wesley, Reading, MA, third edition, 2009.

[100] Victor J. Katz and Karen Hunger Parshall. *Taming the Unknown: A History of Algebra from Antiquity to the Early Twentieth Century.* Princeton University Press, Princeton, NJ, 2014.

[101] Agathe Keller. *Expounding the Mathematical Seed.* Birkhaüser, Basel, 2006.

[102] Patricia Clark Kenschaft. *Change is Possible: Stories of Women and Minorities in Mathematics.* American Mathematical Society, Providence, RI, 2005.

[103] Johannes Kepler. *Optics: Paralipomena to Witelo & Optical Part of Astronomy.* Green Lion Press, Santa Fe, NM, 2000.

[104] Omar Khayyam. *The Algebra of Omar Khayyam, translated by Daoud S. Kasir.* Number 385 in Contributions to Education. Columbia University Teachers College, New York, 1931.

[105] Peggy Aldrich Kidwell. The metric system enters the American classroom: 1790–1890. In Amy Shell-Gellasch and Dick Jardine, editors, *From Calculus to Computers*, pages 229–236. Mathematical Association of America, Washington, DC, 2005.

[106] Jacob Klein. *Greek Mathematical Thought and the Origin of Algebra.* Dover Publications, New York, 1992.

[107] Morris Kline. *Mathematical Thought from Ancient to Modern Times.* Oxford University Press, Oxford and New York, second edition, 1990.

[108] Wilbur R. Knorr. *The Ancient Tradition of Geometric Problems.* Dover Publications, New York, 1993.

[109] Steven Krantz. *Mathematical Apocrypha.* Mathematical Association of America, Washington, DC, 2002.

[110] Steven Krantz. *Mathematical Apocrypha Redux.* Mathematical Association of America, Washington, DC, 2005.

[111] Federica La Nave and Barry Mazur. Reading Bombelli. *The Mathematical Intelligencer*, 24:12–21, 2002.

[112] Phillip H. De Lacy and Benedict Einarson, editors. *Plutarch's Moralia*, volume VII of *Loeb Classical Library*. Harvard University Press, Cambridge, MA, 1959.

[113] R. E. Langer. Josiah Willard Gibbs. *American Mathematical Monthly*, 46:75–84, 1939.

[114] Reinhard Laubenbacher and David Pengelley. *Mathematical Expeditions: Chronicles by the Explorers*. Springer-Verlag, Berlin, Heidelberg, New York, 1999.

[115] Yan Li and Shi Ran Du. *Chinese Mathematics*. Oxford University Press/The Clarendon Press, Oxford and New York, 1987.

[116] Lillian R. Lieber. *Infinity: Beyond the Beyond the Beyond*. Paul Dry Books, New York, 2007. Originally published by Holt, Rinehart and Winston, 1953.

[117] Annette Lykknes, Donald L. Opitz, and Brigitte Van Tiggelen, editors. *For Better or For Worse? Collaborative Couples in the Sciences*. Birkhäuser, Basel, 2012.

[118] Michael S. Mahoney. *The Mathematical Career of Pierre de Fermat, 1601–1665*. Princeton University Press, Princeton, NJ, second edition, 1994.

[119] Eli Maor. *Trigonometric Delights*. Princeton University Press, Princeton, NJ, 1998.

[120] Jean-Claude Martzloff. *A History of Chinese Mathematics*. Springer-Verlag, Berlin, Heidelberg, New York, 1997.

[121] Barry Mazur. *Imagining Numbers (Particularly the Square Root of Minus Fifteen)*. Farrar, Strauss and Giroux, New York, 2002.

[122] Joseph Mazur. *Enlightening Symbols: A Short History of Mathematical Notation and Its Hidden Powers*. Princeton University Press, Princeton, NJ, 2014.

[123] Karl Menninger. *Number Words and Number Symbols: A Cultural History of Numbers*. Dover Publications, New York, 1992.

[124] N. Metropolis, J. Howlett, and Gian-Carlo Rota, editors. *A History of Computing in the Twentieth Century*. Academic Press, New York, 1980.

[125] Henrietta O. Midonick, editor. *The Treasury of Mathematics*. Philosophical Library, Inc., New York, 1965.

[126] Robert Edouard Moritz. *Memorabilia Mathematica*. Mathematical Association of America, Washington, DC, 1993.

[127] C. Morrow and Teri Perl. *Notable Women in Mathematics: A Biographical Dictionary.* Greenwood Press, Westport, CT, 1998.

[128] Margaret A. M. Murray. *Women Becoming Mathematicians: Creating a Professional Identity in post-World War II America.* MIT Press, Cambridge, MA, 2000.

[129] Dora Musielak. *Sophie's Diary: A Mathematical Novel.* Mathematical Association of America, Washington, DC, 2012.

[130] Paul J. Nahin. *An Imaginary Tale: The Story of $\sqrt{-1}$.* Princeton University Press, Princeton, NJ, 1998.

[131] Reviel Netz. *The Shaping of Deduction in Greek Mathematics.* Cambridge University Press, Cambridge, 1999.

[132] Reviel Netz and William Noel. *The Archimedes Codex: Revealing the Secrets of the World's Greatest Palimpsest.* Da Capo Press, 2007.

[133] James R. Newman, editor. *The World of Mathematics.* Dover Publications, New York, 2000. Vols. 1–4.

[134] Deborah Nolan, editor. *Women in Mathematics: Scaling the Heights*, volume 46 of *MAA Notes*. Mathematical Association of America, Washington, DC, 1997.

[135] Apollonius of Perga. *Conics I–IV.* Green Lion Press, Santa Fe, NM, 2013. Books I–III translated by R. Catesby Taliaferro; Book IV translated by Michael N. Fried.

[136] Lynn M. Osen. *Women in Mathematics.* The MIT Press, Cambridge, MA, 1974.

[137] Robert Osserman. *Poetry of the Universe.* Anchor Books, New York, 1996.

[138] Marla Parker, editor. *She Does Math!* Mathematical Association of America, Washington, DC, 1995.

[139] Teri Perl. *Math Equals.* Addison-Wesley, Reading, MA, 1978.

[140] Kim Plofker. *Mathematics in India.* Princeton University Press, Princeton, NJ, 2009.

[141] Walter Prenowitz and Meyer Jordan. *Basic Concepts of Geometry.* Ardsley House, New York, 1989. Originally published by John Wiley & Sons, 1965.

[142] R. Preston. Profile: The Mountains of Pi. *The New Yorker*, pages 36–67, March 2, 1992.

[143] Helena M. Pycior. *Symbols, Impossible Numbers, and Geometric Entanglements: British Algebra Through the Commentaries on Newton's Universal Arithmetick*. Cambridge University Press, Cambridge, 1997.

[144] R. Rashed, editor. *Encyclopaedia of the History of Arabic Sciences*. Routledge, London and New York, 1996.

[145] H. L. Resnikoff and R. O. Wells, Jr. *Mathematics in Civilization*. Dover Publications, New York, 1984.

[146] Eleanor Robson. *Mesopotamian Mathematics, 2100–1600 BC: Technical Constants in Bureaucracy and Education*. Clarendon Press, Oxford, 1999.

[147] Eleanor Robson. *Mathematics in Ancient Iraq: A Social History*. Princeton University Press, Princeton, NJ, 2008.

[148] John J. Roche. *The Mathematics of Measurement: A Critical History*. Athlone Press/Springer-Verlag, London/New York, 1998.

[149] Margaret Rossiter. *Women Scientists in America*. Johns Hopkins University Press, Baltimore, 1982.

[150] George Saliba. *Islamic Science and the Making of the European Renaissance*. MIT Press, Cambridge, MA, 2007.

[151] David Salsburg. *The Lady Tasting Tea*. W. H. Freeman, New York, 2001.

[152] Rosemary Schmalz. *Out of the Mouths of Mathematicians*. Mathematical Association of America, Washington, DC, 1993.

[153] Denise Schmandt-Besserat. Oneness, twoness, threeness. In Swetz [170], pages 45–51.

[154] Denise Schmandt-Besserat. *The History of Counting*. Morrow Junior Books, New York, 1999. Illustrated by Michael Hays.

[155] Randy K. Schwartz. Issues in the origin and development of *Hisab al-Khata'ayn* (calculation by double false position). In *Actes du Huitième Colloque Maghrébin sur l'Histoire des Mathématiques Arabes, Tunis, les 18–19–20 Decembre 2004*, Tunis, 2006. Tunisian Association of Mathematical Sciences.

[156] Randy K. Schwartz. Adapting the medieval "rule of false position" to the modern classroom. In Jardine and Shell-Gellasch [93], pages 29–37.

[157] Helaine Selin and Ubiratan D'Ambrosio, editors. *Mathematics Across Cultures: The History of Non-Western Mathematics.* Kluwer Academic, Dordrecht, Boston, London, 2000.

[158] Jacques Sesiano. *An Introduction to the History of Algebra: Solving Equations from Mesopotamian Times to the Renaissance.* American Mathematical Society, Providence, RI, 2009. Translated by Anna Pierrehumbert.

[159] Kangshen Shen, John N. Crossley, and Anthony W.-C. Lun. *The Nine Chapters on the Mathematical Art: Companion and Commentary.* Oxford University Press, Oxford and New York, 1999.

[160] John R. Silvester. Decimal déjà vu. *The Mathematical Gazette,* 83:453–463, 1999.

[161] Simon Singh. *Fermat's Enigma.* Walker and Company, New York, 1997.

[162] David Singmaster. Chronology of recreational mathematics. Online at anduin.eldar.org/~problemi/singmast/recchron.html.

[163] David Singmaster. Some early sources in recreational mathematics. In Hay [85], pages 195–208.

[164] David Eugene Smith. *History of Mathematics.* Dover Publications, New York, 1958. Vols. 1 and 2.

[165] David Eugene Smith. *A Source Book in Mathematics.* Dover Publications, New York, 1959.

[166] Dava Sobel. *Longitude: The True Story of a Lone Genius Who Solved the Greatest Scientific Problem of His Time.* Penguin Books, New York, 1995.

[167] Jacqueline Stedall. *Mathematics Emerging: A Sourcebook 1540–1900.* Oxford University Press, Oxford, 2008.

[168] Stephen M. Stigler. *The History of Statistics.* Harvard University Press, Cambridge, MA, 1986.

[169] Frank Swetz, John Fauvel, Otto Bekken, Bengt Johansson, and Victor Katz, editors. *Learn from the Masters*. Mathematical Association of America, Washington, DC, 1995.

[170] Frank J. Swetz, editor. *From Five Fingers to Infinity*. Open Court, Chicago, 1994.

[171] Frank J. Swetz. *The Search for Certainty: A Journey Through the History of Mathematics from 1800–2000*. Dover, New York, 2012.

[172] Frank J. Swetz. *The European Mathematical Awakening: A Journey Through the History of Mathematics from 1000 to 1800*. Dover, New York, 2013.

[173] J. J. Sylvester. *The Collected Mathematical Papers of James Joseph Sylvester*. AMS Chelsea, Providence, RI, 1973.

[174] Renate Tobies. *Iris Runge: A Life at the Crossroads of Mathematics, Science, and Industry*. Birkhäuser, Basel, 2013.

[175] Glen Van Brummelen. *The Mathematics of the Heavens and the Earth: The Early History of Trigonometry*. Princeton University Press, Princeton, NJ, 2009.

[176] D. T. Whiteside, editor. *The Mathematical Papers of Isaac Newton*. Cambridge University Press, Cambridge, 1972.

[177] Robin J. Wilson. *Stamping Through Mathematics*. Springer-Verlag, Berlin, Heidelberg, New York, 2001.

[178] Benjamin H. Yandell. *The Honors Class: Hilbert's Problems and Their Solvers*. A K Peters, Natick, MA, 2001.

Index